崧燁文化

德 著

Arduino程式教學
(顯示模組篇)

Arduino Programming (Display Modules)

自序

    Arduino 系列的書出版至今，已經過三年，出書量也近八十本大關，當初出版電子書是希望能夠在教育界開一門 Maker 自造者相關的課程，沒想到一寫就已過三年，繁簡體加起來的出版數也已也近八十本的量，這些書都是我學習當一個 Maker 累積下來的成果。

    這本書可以說是我的書另一個里程碑，很久以前，這個系列開始以駭客的觀點為主，希望 Maker 可以擁有駭客的觀點、技術、能力，駭入每一個產品設計思維，並且成功的重製、開發、超越原有的產品設計，這才是一位對社會有貢獻的『駭客』。

    如許多學習程式設計的學子，為了最新的科技潮流，使用著最新的科技工具與軟體元件，當他們面對許多原有的軟體元件沒有支持的需求或軟體架構下沒有直接直持的開發工具，此時就產生了莫大的開發瓶頸，這些都是為了追求最新的科技技術而忘卻了學習原有基礎科技訓練所致。

    筆著鑒於這樣的困境，思考著『如何駭入眾人現有知識寶庫轉換為我的知識』的思維，如果我們可以駭入產品結構與設計思維，那麼了解產品的機構運作原理與方法就不是一件難事了。更進一步我們可以將原有產品改造、升級、創新，並可以將學習到的技術運用其他技術或新技術領域，透過這樣學習思維與方法，可以更快速的掌握研發與製造的核心技術，相信這樣的學習方式，會比起在已建構好的開發模組或學習套件中學習某個新技術或原理，來的更踏實的多。

    目前許多學子在學習程式設計之時，恐怕最不能了解的問題是，我為何要寫九九乘法表、為何要寫遞迴程式，為何要寫成函式型式…等等疑問，只因為在學校的學子，學習程式是為了可以了解『撰寫程式』的邏輯，並訓練且建立如何運用程式邏輯的能力，解譯現實中面對的問題。然而現實中的問題往往太過於複雜，授課的老師無法有多餘的時間與資源去解釋現實中複雜問題，期望能將現實中複雜問題淬鍊成邏輯上的思路，加以訓練學生其解題思路，但是眾多學子宥於現實問題的困惑，無法單純用純粹的解題思路來進行學習與訓練，反而以現實中的複雜來反駁老師教學太過學理，沒有實務上的應用為由，拒絕深入學習，這樣的情形，反而自己造成了學習上的障礙。

本系列的書籍，針對目前學習上的盲點，希望讀者當一位產品駭客，將現有產品的產品透過逆向工程的手法，進而了解核心控制系統之軟硬體，再透過簡單易學的 Arduino 單晶片與 C 語言，重新開發出原有產品，進而改進、加強、創新其原有產品固有思維與架構。如此一來，因為學子們進行『重新開發產品』過程之中，可以很有把握的了解自己正在進行什麼，對於學習過程之中，透過實務需求導引著開發過程，可以讓學子們讓實務產出與邏輯化思考產生關連，如此可以一掃過去陰霾，更踏實的進行學習。

這三年多以來的經驗分享，逐漸在這群學子身上看到發芽，開始成長，覺得 Maker 的教育方式，極有可能在未來成為教育的主流，相信我每日、每月、每年不斷的努力之下，未來 Maker 的教育、推廣、普及、成熟將指日可待。

最後，請大家可以加入 Maker 的 Open Knowledge 的行列。

曹永忠 於貓咪樂園

自序

記得自己在大學資訊工程系修習電子電路實驗的時候，自己對於設計與製作電路板是一點興趣也沒有，然後又沒有天分，所以那是苦不堪言的一堂課，還好當年有我同組的好同學，努力的照顧我，命令我做這做那，我不會的他就自己做，如此讓我解決了資訊工程學系課程中，我最不擅長的課。

當時資訊工程學系對於設計電子電路課程，大多數都是專攻軟體的學生去修習時，系上的用意應該是要大家軟硬兼修，尤其是在台灣這個大部分是硬體為主的產業環境，但是對於一個軟體設計，但是缺乏硬體專業訓練，或是對於眾多機械機構與機電整合原理不太有概念的人，在理解現代的許多機電整合設計時，學習上都會有很多的困擾與障礙，因為專精於軟體設計的人，不一定能很容易就懂機電控制設計與機電整合。懂得機電控制的人，也不一定知道軟體該如何運作，不同的機電控制或是軟體開發常常都會有不同的解決方法。

除非您很有各方面的天賦，或是在學校巧遇名師教導，否則通常不太容易能在機電控制與機電整合這方面自我學習，進而成為專業人員。

而自從有了 Arduino 這個平台後，上述的困擾就大部分迎刃而解了，因為 Arduino 這個平台讓你可以以不變應萬變，用一致性的平台，來做很多機電控制、機電整合學習，進而將軟體開發整合到機構設計之中，在這個機械、電子、電機、資訊、工程等整合領域，不失為一個很大的福音，尤其在創意掛帥的年代，能夠自己創新想法，從 Original Idea 到產品開發與整合能夠自己獨立完整設計出來，自己就能夠更容易完全了解與掌握核心技術與產業技術，整個開發過程必定可以提供思維上與實務上更多的收穫。

Arduino 平台引進台灣自今，雖然越來越多的書籍出版，但是從設計、開發、製作出一個完整產品並解析產品設計思維，這樣產品開發的書籍仍然鮮見，尤其是能夠從頭到尾，利用範例與理論解釋並重，完完整整的解說如何用 Arduino 設計出一個完整產品，介紹開發過程中，機電控制與軟體整合相關技術與範例，如此的書籍更是付之闕如。永忠、英德兄與敝人計畫撰寫 Maker 系列，就是基於這樣對市場需要的觀察，開發出這樣的書籍。

作者出版了許多的 Arduino 系列的書籍，深深覺的，基礎乃是最根本的實力，所以回到最基礎的地方，希望透過最基本的程式設計教學，來提供眾多的 Makers 在入門 Arduino 時，如何開始，如何攥寫自己的程式，進而介紹不同的週邊模組，主要的目的是希望學子可以學到如何使用這些週邊模組來設計程式，期望在未來產品開發時，可以更得心應手的使用這些週邊模組與感測器，更快將自己的想法實現，希望讀者可以了解與學習到作者寫書的初衷。

許智誠　　於中壢雙連坡中央大學　管理學院

自序

隨著資通技術(ICT)的進步與普及,取得資料不僅方便快速,傳播資訊的管道也多樣化與便利。然而,在網路搜尋到的資料卻越來越巨量,如何將在眾多的資料之中篩選出正確的資訊,進而萃取出您要的知識?如何獲得同時具廣度與深度的知識?如何一次就獲得最正確的知識?相信這些都是大家共同思考的問題。

為了解決這些困惱大家的問題,永忠、智誠兄與敝人計畫製作一系列「Maker系列」書籍來傳遞兼具廣度與深度的軟體開發知識,希望讀者能利用這些書籍迅速掌握正確知識。首先規劃「以一個 Maker 的觀點,找尋所有可用資源並整合相關技術,透過創意與逆向工程的技法進行設計與開發」的系列書籍,運用現有的產品或零件,透過駭入產品的逆向工程的手法,拆解後並重製其控制核心,並使用 Arduino相關技術進行產品設計與開發等過程,讓電子、機械、電機、控制、軟體、工程進行跨領域的整合。

近年來 Arduino 異軍突起,在許多大學,甚至高中職、國中,甚至許多出社會的工程達人,都以 Arduino 為單晶片控制裝置,整合許多感測器、馬達、動力機構、手機、平板...等,開發出許多具創意的互動產品與數位藝術。由於 Arduino 的簡單、易用、價格合理、資源眾多,許多大專院校及社團都推出相關課程與研習機會來學習與推廣。

以往介紹 ICT 技術的書籍大部份以理論開始、為了深化開發與專業技術,往往忘記這些產品產品開發背後所需要的背景、動機、需求、環境因素等,讓讀者在學習之間,不容易了解當初開發這些產品的原始創意與想法,基於這樣的原因,一般人學起來特別感到吃力與迷惘。

本書為了讀者能夠深入了解產品開發的背景,本系列整合 Maker 自造者的觀念與創意發想,深入產品技術核心,進而開發產品,只要讀者跟著本書一步一步研習與實作,在完成之際,回頭思考,就很容易了解開發產品的整體思維。透過這樣的思路,讀者就可以輕易地轉移學習經驗至其他相關的產品實作上。

所以本書是能夠自修的書，讀完後不僅能依據書本的實作說明準備材料來製作，盡情享受 DIY(Do It Yourself)的樂趣，還能了解其原理並推展至其他應用。有興趣的讀者可再利用書後的參考文獻繼續研讀相關資料。

本書的發行有新的創舉，就是以電子書型式發行，在國家圖書館 (http://www.ncl.edu.tw/)、國立公共資訊圖書館 National Library of Public Information(http://www.nlpi.edu.tw/)、台灣雲端圖庫(http://www.ebookservice.tw/)等都可以免費借閱與閱讀，如要購買的讀者也可以到許多電子書網路商城、Google Books 與 Google Play 都可以購買之後下載與閱讀。希望讀者能珍惜機會閱讀及學習，繼續將知識與資訊傳播出去，讓有興趣的眾人都受益。希望這個拋磚引玉的舉動能讓更多人響應與跟進，一起共襄盛舉。

本書可能還有不盡完美之處，非常歡迎您的指教與建議。近期還將推出其他 Arduino 相關應用與實作的書籍，敬請期待。

最後，請您立刻行動翻書閱讀。

蔡英德 於台中沙鹿靜宜大學主顧樓

目 錄

Maker 系列

　　本書是『Arduino 程式教學』的第六本書，主要是給讀者熟悉 Arduino 的視覺
輸出模組：顯示模組。Arduino 開發板最強大的不只是它的簡單易學的開發工具，
最強大的是它封富的周邊模組與簡單易學的模組函式庫，幾乎 Maker 想到的東西，
都有廠商或 Maker 開發它的周邊模組，透過這些周邊模組，Maker 可以輕易的將想
要完成的東西用堆積木的方式快速建立，而且最強大的是這些周邊模組都有對應的
函式庫，讓 Maker 不需要具有深厚的電子、電機與電路能力，就可以輕易駕御這些
模組。

　　所以本書要介紹市面上最常見、最受歡迎與使用的顯示模組，讓讀者可以輕鬆
學會這些常用模組的使用方法，進而提升各位 Maker 的實力。

CHAPTER

Arduino 簡介

Massimo Banzi 之前是義大利 Ivrea 一家高科技設計學校的老師，他的學生們經常抱怨找不到便宜好用的微處理機控制器。西元 2005 年， Massimo Banzi 跟 David Cuartielles 討論了這個問題，David Cuartielles 是一個西班牙籍晶片工程師，當時是這所學校的訪問學者。兩人討論之後，決定自己設計電路板，並引入了 Banzi 的學生 David Mellis 為電路板設計開發用的語言。兩天以後，David Mellis 就寫出了程式碼。又過了幾天，電路板就完工了。於是他們將這塊電路板命名為『Arduino』。

當初 Arduino 設計的觀點，就是希望針對『不懂電腦語言的族群』，也能用 Arduino 做出很酷的東西，例如：對感測器作出回應、閃爍燈光、控制馬達…等等。

隨後 Banzi，Cuartielles，和 Mellis 把設計圖放到了網際網路上。他們保持設計的開放源碼(Open Source)理念，因為版權法可以監管開放原始碼軟體，卻很難用在硬體上，他們決定採用創用 CC 許可(Creative_Commons, 2013)。

創用 CC(Creative_Commons, 2013)是為保護開放版權行為而出現的類似 GPL[1] 的一種許可（license），來自於自由軟體[2]基金會 (Free Software Foundation) 的 GNU 通用公共授權條款 (GNU GPL)：在創用 CC 許可下，任何人都被允許生產電路板的複製品，且還能重新設計，甚至銷售原設計的複製品。你還不需要付版稅，甚至不用取得 Arduino 團隊的許可。

然而，如果你重新散佈了引用設計，你必須在其產品中註解說明原始 Arduino 團隊的貢獻。如果你調整或改動了電路板，你的最新設計必須使用相同或類似的創用 CC 許可，以保證新版本的 Arduino 電路板也會一樣的自由和開放。

---

[1] GNU 通用公眾授權條款（英語：GNU General Public License，簡稱 GNU GPL 或 GPL），是一個廣泛被使用的自由軟體授權條款，最初由理察·斯托曼為 GNU 計劃而撰寫。

[2] 「自由軟體」指尊重使用者及社群自由的軟體。簡單來說使用者可以自由運行、複製、發佈、學習、修改及改良軟體。他們有操控軟體用途的權利。

唯一被保留的只有 Arduino 這個名字：『Arduino』已被註冊成了商標[3]『Arduino®』。如果有人想用這個名字賣電路板，那他們可能必須付一點商標費用給 『Arduino®』 (Arduino, 2013)的核心開發團隊成員。

『Arduino®』的核心開發團隊成員包括：Massimo Banzi，David Cuartielles，Tom Igoe，Gianluca Martino，David Mellis 和 Nicholas Zambetti。(Arduino, 2013)，若讀者有任何不懂 Arduino 的地方，都可以訪問 Arduino 官方網站：http://www.arduino.cc/

『Arduino®』，是一個開放原始碼的單晶片控制器，它使用了 Atmel AVR 單晶片 (Atmel_Corporation, 2013)，採用了基於開放原始碼的軟硬體平台，構建於開放原始碼 Simple I/O 介面版，並且具有使用類似 Java，C 語言的 Processing[4]/Wiring 開發環境(B. F. a. C. Reas, 2013; C. Reas & Fry, 2007, 2010)。Processing 由 MIT 媒體實驗室美學與計算小組 (Aesthetics & Computation Group) 的 Ben Fry(http://benfry.com/)和 Casey Reas 發明，Processing 已經有許多的 Open Source 的社群所提倡，對資訊科技的發展是一個非常大的貢獻。

讓您可以快速使用 Arduino 語言作出互動作品，Arduino 可以使用開發完成的電子元件：例如 Switch、感測器、其他控制器件、LED、步進馬達、其他輸出裝置…等。Arduino 開發 IDE 介面基於開放原始碼，可以讓您免費下載使用，開發出更多令人驚豔的互動作品(Banzi, 2009) 。

---

[3] 商標註冊人享有商標的專用權，也有權許可他人使用商標以獲取報酬。各國對商標權的保護期限長短不一，但期滿之後，只要另外繳付費用，即可對商標予以續展，次數不限。

[4] Processing 是一個 Open Source 的程式語言及開發環境，提供給那些想要對影像、動畫、聲音進行程式處理的工作者。此外，學生、藝術家、設計師、建築師、研究員以及有興趣的人，也可以用來學習，開發原型及製作

什麼是 Arduino

- Arduino 是基於開放原碼精神的一個開放硬體平臺，其語言和開發環境都很簡單。讓您可以使用它快速做出有趣的東西。

- 它是一個能夠用來感應和控制現實物理世界的一套工具，也提供一套設計程式的 IDE 開發環境，並可以免費下載

- Arduino 可以用來開發互動產品，比如它可以讀取大量的開關和感測器信號，並且可以控制各式各樣的電燈、電機和其他物理設備。也可以在運行時和你電腦中運行的程式（例如：Flash，Processing，MaxMSP）進行通訊。

Arduino 特色

- 開放原始碼的電路圖設計，程式開發介面

- http://www.arduino.cc/免費下載，也可依需求自己修改!!

- Arduino 可使用 ISCP 線上燒入器，自我將新的 IC 晶片燒入「bootloader」(http://arduino.cc/en/Hacking/Bootloader?from=Main.Bootloader) 。

- 可依據官方電路圖(http://www.arduino.cc/)，簡化 Arduino 模組，完成獨立運作的微處理機控制模組

- 感測器可簡單連接各式各樣的電子元件 (紅外線,超音波,熱敏電阻,光敏電阻,伺服馬達,…等)

- 支援多樣的互動程式程式開發工具

- 使用低價格的微處理控制器(ATMEGA8-16)

- USB 介面，不需外接電源。另外有提供 9VDC 輸入

- 應用方面，利用 Arduino，突破以往只能使用滑鼠，鍵盤，CCD 等輸入的裝置的互動內容，可以更簡單地達成單人或多人遊戲互動

Arduino 硬體-Yun 雲

Arduino Yún 是 Arduino 最新的開發板，這是 Arduino 公司 Wi-Fi 生產線的首項產品。把 Arduino Leonardo 開發板的功能和以 Linux 為基礎的無線路由器合為一體。基本上是將 WiFi Linux 板內建至 Leonardo（ATmega32U4），同時藉由 Linino（MIPS Linux 改良版）對應 XML 等純文字（Text-Based）格式，並進行其它的 HTTP 交易事務。

系統規格

AVR Arduino microcontroller

- 控制晶片：ATmega32u4
- 工作電壓：5V
- 輸入電壓：5V
- Digital I/O Pins：20
- PWM Channels：7
- Analog Input Channels：12
- DC Current per I/O Pin：40 mA
- DC Current for 3.3V Pin：50 mA
- Flash Memory：32 KB (of which 4 KB used by bootloader)
- SRAM：2.5 KB
- EEPROM：1 KB
- Clock Speed：16 MHz

Linux microprocessor

- 處理器：Atheros AR9331
- Architecture：MIPS @400MHz
- 工作電壓：3.3V
- Ethernet：IEEE 802.3 10/100Mbit/s
- WiFi：IEEE 802.11b/g/n
- USB Type-A 2.0 Host/Device

- Card Reader：Micro-SD only
- RAM：64 MB DDR2
- Flash Memory：32 MB
- PoE compatible 802.3af card support

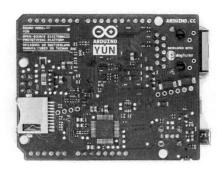

圖 1 Arduino Yun 開發板外觀圖

如下圖所示，Arduino Yun 開發板上有 3 個重置鍵：

- 如下圖所示，左上標示為 32U4 RST 的重置鍵，重置 ATmega32U4 這顆微控制器。

- 如下圖所示，右下標示為 Yún RST 的重置鍵，重置 AR9331，重新啟動 Linux 系統（Linino），記憶體中的東西全部不見，執行中的程式也會終止。

- 如下圖所示。左下標示為 WLAN RST 的重置鍵，有兩個作用，第一是將 WiFi 組態重置回工廠設定值，會讓 WiFi 晶片進入 AP（access point）模式，IP 是 192.168.240.1，分享出來的網路名稱是「Arduino Yun-XXXXXXXXXXXX」，其中 X 是 WiFi 無線網路卡的 MAC 位址，按著此重置鍵不放、持續 5 秒，即可進入 WiFi 組態重置模式。第二個作用是將 Linux 映像檔重置回工廠預設的映像檔內容，必須按著重置鍵不放持續 30 秒，這麼一來，儲存在板子裡的快閃記憶體（與 AR9331 連接）

的檔案，通通都會消失。

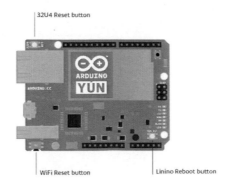

圖 2 Arduino Yun 開發板重置鍵一覽圖

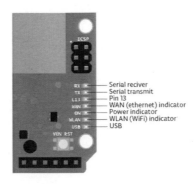

圖 3 Arduino Yun 開發板燈號圖

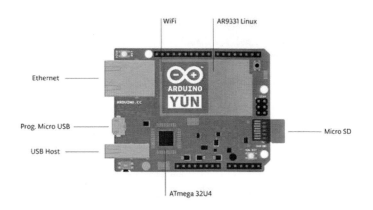

圖表 4 Arduino Yun 開發板接口一覽圖

  Arduino Yun 開發板是一款設備齊全的有線、無線網路單晶片處理器，此處理器可簡化網際網路連接的複雜執行過程。由下圖所示，可以了解為了將原有 Arduino架構保留，Arduino Yun 開發板採用橋接器的架構來括充有線/無線網路的功能，讓使用者可以輕易整合，也可以讓複雜的網路程式變得容易開發與維護。

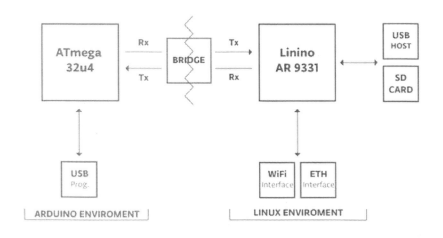

圖 5 Arduino Yun 開發板運作架構圖

  讀者可以由下面程式看到，只要使用『Console.h』的函式庫(內建於 Sketch IDE環境)，就可以簡單的與另外的架構程式溝通。

```
#include <Console.h>

const int ledPin = 13; // the pin that the LED is attached to
int incomingByte;        // a variable to read incoming serial data into

void setup() {
  // initialize serial communication:
  Bridge.begin();
  Console.begin();

  while (!Console){
    ; // wait for Console port to connect.
  }
  Console.println("You're connected to the Console!!!!");
  // initialize the LED pin as an output:
  pinMode(ledPin, OUTPUT);
}

void loop() {
  // see if there's incoming serial data:
  if (Console.available() > 0) {
    // read the oldest byte in the serial buffer:
    incomingByte = Console.read();
    // if it's a capital H (ASCII 72), turn on the LED:
    if (incomingByte == 'H') {
      digitalWrite(ledPin, HIGH);
    }
    // if it's an L (ASCII 76) turn off the LED:
    if (incomingByte == 'L') {
        digitalWrite(ledPin, LOW);
      }
    }
    delay(100);
  }
```

Arduino 硬體-Duemilanove

Arduino Duemilanove 使用 AVR Mega168 為微處理晶片，是一件功能完備的單晶片開發板，Duemilanove 特色為：(a).開放原始碼的電路圖設計，(b).程序開發免費下載，(c).提供原始碼可提供使用者修改，(d).使用低價格的微處理控制器(ATmega168)，(e).採用 USB 供電，不需外接電源，(f).可以使用外部 9VDC 輸入，(g).支持 ISP 直接線上燒錄，(h).可使用 bootloader 燒入 ATmega8 或 ATmega168 單晶片。

系統規格

- 主要溝通介面:USB
- 核心: ATMEGA328
- 自動判斷並選擇供電方式（USB/外部供電）
- 控制器核心：ATmega328
- 控制電壓：5V
- 建議輸入電(recommended)：7-12 V
- 最大輸入電壓 (limits)：6-20 V
- 數位 I/O Pins：14 (of which 6 provide PWM output)
- 類比輸入 Pins：6 組
- DC Current per I/O Pin：40 mA
- DC Current for 3.3V Pin：50 mA
- Flash Memory：32 KB (of which 2 KB used by bootloader)
- SRAM：2 KB
- EEPROM：1 KB
- Clock Speed：16 MHz

具有 bootloader⁵能夠燒入程式而不需經過其他外部電路。此版本設計了『自動

---

⁵ 啟動程式（boot loader）位於電腦或其他計算機應用上，是指引導操作系統啟動的程式。

回復保險絲[6]』，在 Arduino 開發板搭載太多的設備或電路短路時能有效保護 Arduino 開發板的 USB 通訊埠，同時也保護了您的電腦，並且故障排除後能自動恢復正常。

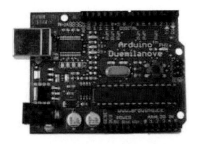 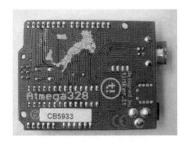

圖 6 Arduino Duemilanove 開發板外觀圖

Arduino 硬體-UNO

UNO 的處理器核心是 ATmega328，使用 ATMega 8U2 來當作 USB-對序列通訊，並多了一組 ICSP 給 MEGA8U2 使用：未來使用者可以自行撰寫內部的程式~ 也因為捨棄 FTDI USB 晶片~ Arduino 開發板需要多一顆穩壓 IC 來提供 3.3V 的電源。

Arduino UNO 是 Arduino USB 介面系列的最新版本，作為 Arduino 平臺的參考標準範本： 同時具有 14 路數位輸入/輸出口（其中 6 路可作為 PWM 輸出），6 路模擬輸入， 一個 16MHz 晶體振盪器，一個 USB 口，一個電源插座，一個 ICSP header 和一個重定按鈕。

UNO 目前已經發佈到第三版，與前兩版相比有以下新的特點： (a).在 AREF 處增加了兩個管腳 SDA 和 SCL，(b).支援 I2C 介面，(c).增加 IOREF 和一個預留管腳，將來擴展板將能相容 5V 和 3.3V 核心板，(d).改進了 Reset 重置的電路設計，

---

[6]自恢復保險絲是一種過流電子保護元件，採用高分子有機聚合物在高壓、高溫，硫化反應的條件下，攙加導電粒子材料後，經過特殊的生產方法製造而成。Ps. PPTC(PolyerPositiveTemperature Coefficent)也叫自恢復保險絲。嚴格意義講：PPTC 不是自恢復保險絲，ResettableFuse 才是自恢復保險絲。

(e).USB 介面晶片由 ATmega16U2 替代了 ATmega8U2。

系統規格

- 控制器核心：ATmega328
- 控制電壓：5V
- 建議輸入電(recommended)：7-12 V
- 最大輸入電壓 (limits)：6-20 V
- 數位 I/O Pins：14 (of which 6 provide PWM output)
- 類比輸入 Pins：6 組
- DC Current per I/O Pin：40 mA
- DC Current for 3.3V Pin：50 mA
- Flash Memory：32 KB (of which 0.5 KB used by bootloader)
- SRAM：2 KB
- EEPROM：1 KB
- Clock Speed：16 MHz

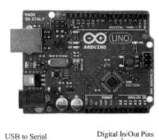

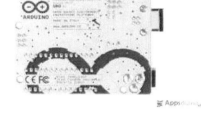

圖 7 Arduino UNO 開發板外觀圖

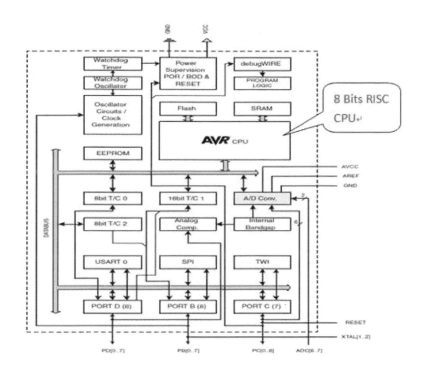

圖 8 Arduino UNO 核心晶片 Atmega328P 架構圖

Arduino 硬體-Mega 2560

可以說是 Arduino 巨大版： Arduino Mega2560 REV3 是 Arduino 官方最新推出的 MEGA 版本。功能與 MEGA1280 幾乎是一模一樣，主要的不同在於 Flash 容量從 128KB 提升到 256KB，比原來的 Atmega1280 大。

Arduino Mega2560 是一塊以 ATmega2560 為核心的微控制器開發板，本身具有54 組數位 I/O input/output 端（其中 14 組可做 PWM 輸出），16 組模擬比輸入端，4組 UART（hardware serial ports），使用 16 MHz crystal oscillator。由於具有bootloader，因此能夠通過 USB 直接下載程式而不需經過其他外部燒入器。供電部份可選擇由 USB 直接提供電源，或者使用 AC-to-DC adapter 及電池作為外部供電。

由於開放原代碼，以及使用 Java 概念（跨平臺）的 C 語言開發環境，讓 Arduino

的周邊模組以及應用迅速的成長。而吸引 Artist 使用 Arduino 的主要原因是可以快速使用 Arduino 語言與 Flash 或 Processing...等軟體通訊,作出多媒體互動作品。Arduino 開發 IDE 介面基於開放原代碼原則,可以讓您免費下載使用於專題製作、學校教學、電機控制、互動作品等等。

電源設計

Arduino Mega2560 的供電系統有兩種選擇,USB 直接供電或外部供電。電源供應的選擇將會自動切換。外部供電可選擇 AC-to-DC adapter 或者電池,此控制板的極限電壓範圍為 6V~12V,但倘若提供的電壓小於 6V,I/O 口有可能無法提供到 5V 的電壓,因此會出現不穩定;倘若提供的電壓大於 12V,穩壓裝置則會有可能發生過熱保護,更有可能損壞 Arduino MEGA2560。因此建議的操作供電為 6.5~12V,推薦電源為 7.5V 或 9V。

系統規格

- 控制器核心:ATmega2560
- 控制電壓:5V
- 建議輸入電(recommended):7-12 V
- 最大輸入電壓 (limits):6-20 V
- 數位 I/O Pins:54 (of which 14 provide PWM output)
- UART:4 組
- 類比輸入 Pins:16 組
- DC Current per I/O Pin:40 mA
- DC Current for 3.3V Pin:50 mA
- Flash Memory:256 KB of which 8 KB used by bootloader
- SRAM:8 KB
- EEPROM:4 KB
- Clock Speed:16 MHz

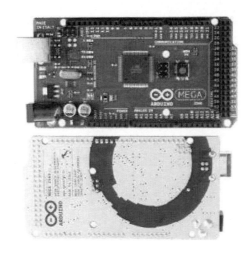

圖 9 Arduino Mega2560 開發板外觀圖

Arduino 硬體- Arduino Pro Mini 控制器

可以說是 Arduino 小型版： Pro Mini 使用 ATMEGA328，與 Arduino Duemilanove 一樣為 5V 並使用 16MHz bootloader，因此在使用 Arduino IDE 時必須選擇
"ArduinoDuemilanove 。

Arduino Pro Mini 控制器為模組大廠 Sparkfun(https://www.sparkfun.com/)依據
Arduino 概念所推出的控制器。藍底 PCB 板以及 0.8mm 的厚度，完全使用 SMD 元
件，讓人看一眼就想馬上知道它有何強大功能。

而 Arduino Pro Mini 與 Arduino Mini 的差異在於，Pro Mini 提供自動 RESET，
使用連接器時只要接上 DTR 腳位與 GRN 腳位，即具備 Autoreset 功能。 而 Pro Mini
與 Duemilanove 的差異點在於 Pro Mini 本身不具備與電腦端相連的轉接器，例如
USB 介面或者 RS232 介面，本身只提供 TTL 準位的 TX、RX 訊號輸出。這樣的
設計較不適合初學者，初學者的入門 建議還是使用 Arduino Duemilanove。

對於熟悉 Arduino 的使用者，可以利用 Pro Mini 為你節省不少成本與體積，你
只需準備一組習慣使用的轉接器，如 UsbtoTTL 轉接器_5V，就可重複使用。

系統規格

- 不包含 USB 連接器以及 USB 轉 TTL 訊號晶片
- 支援 Auto-reset
- ATMEGA328 使用電壓 5V / 頻率 16MHz (external resonator _0.5% tolerance)
- 具 5V 穩壓裝置
- 最大電流 150mA
- 具過電流保護裝置
- 容忍電壓：5-12V
- 內嵌 電源 LED 與狀態 LED
- 尺寸：0.7x1.3" (18x33mm)
- 重量：1.8g
- Arduino 所有特色皆可使用：

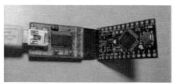

圖 10 Arduino Pro Mini 控制器開發板外觀圖

Arduino 硬體- Arduino ATtiny85 控制器

可以說是 Arduino 超微版： Arduino ATtiny85 是 Atmel Corporation 宣布其低功耗的 ATtiny 10/20/40 微控制器 (MCU) 系列，針對按鍵、滑塊和滑輪等觸控感應應用予以優化。這些元件包括了 AVR MCU 及其專利的低功耗 picoPower 技術，是對成本敏感的工業和消費電子市場上多種應用，如汽車控制板、LCD 電視和顯示器、筆記本電腦、手機等的理想選擇。

ATtiny MCU 系列介紹

Atmel Corporation 設計的 ATtiny 新型單晶片有 AVR 微處理機大部份的功能，以包括 1KB 至 4KB 的 Flash Memory，帶有 32 KB 至 256 KB 的 SRAM。

此外，這些元件支持 SPI 和 TWI (具備 I2C-兼容性) 通訊，提供最高靈活性和 1.8V 至 5.5V 的工作電壓。ATtinyAVR 使用 Atmel Corporation 獨有專利的 picoPower 技術，耗電極低。通過軟件控制系統時鐘頻率，取得系統性能與耗電之間的最佳平衡，同時也得到了廣泛應用。

系統規格

- 採用 ATMEL TINY85 晶片
- 支持 Arduino IDE 1.0+
- USB 供電, 或 7~35V 外部供電
- 共 6 個 I/O 可以用

ATTiny25/45/85/13

D5(A0)/RESET/ADC0/PB5	1		8	VCC
D3(A3)/XTAL1/ADC3/PB3*	2		7	PB2/SCL/SCK/D2(A1)
D4(A2)/XTAL2/ADC2/PB4*	3		6	PB1*/MISO/D1
GND	4		5	PB0*/MOSI/SDA/AREF/D0

NOTES:
- Arduino pins for ATTinyX313/X5/X4 are from the arduino-tiny project
- PWM pins are marked with (*)

*ATTiny13 has PWM only on PB0 and PB1
* No AREF on ATtiny13
* PB3 and PB4 share the same timer

圖 11 Arduino ATtiny85 控制器外觀圖

Arduino 硬體- Arduino LilyPad 控制器

可以說是 Arduino 微小版：Arduino LilyPad 為可穿戴的電子紡織科技由 Leah Buechley 開發及 Leah 及 SparkFun 設計。每一個 LilyPad 設計都有很大的連接點可以縫在衣服上。多種的輸出，輸入，電源，及感測板可以通用，而且還可以水洗。

Arduino LilyPads 主機板的設計包含 ATmega328P(bootloader) 及最低限度的外部元件來維持其簡單小巧特性，可以利用 2-5V 的電壓。 還有加上重置按鈕可以更容易的攥寫程式，Arduino LilyPad 這是一款真正有藝術氣質的產品，很漂亮的造型，當初設計時主要目的就是讓從事服裝設計之類工作的設計師和造型設計師，它可以使用導電線或普通線縫在衣服或布料上, Arduino LilyPad 每個接腳上的小洞大到足夠縫紉針輕鬆穿過。如果用導電線縫紉的話,既可以起到固定的 作用,又可以起到傳導的作用。比起普通的 Arduino 版相比，Arduino LilyPad 相對比較脆弱，比較容易損壞,但它的功能基本都保留了下來, Arduino LilyPad 版子它沒有 USB 介面,所以 Arduino LilyPad 連接電腦或燒寫程式時同 Arduino mini 一樣需要一個 USB 或 RS232 轉換成 TTL 的轉接腳。

系統規格

- 微控制器：ATmega328V
- 工作電壓：2.7-5.5V
- 輸入電壓：2.7-5.5V
- 數位 I / O 接腳：14（其中 6 提供 PWM 輸出）
- 類比輸入接腳：6
- 每個 I / O 接腳的直流電流：40mA
- 快閃記憶體：16 KB（其中 2 KB 使用引導程序）
- SRAM：1 KB
- EEPROM：512k
- 時鐘速度：8 MHz

圖 12 Arduino LilyPad 控制器外觀圖

Arduino 硬體- Arduino Esplora 控制器

Arduino Esplora 可是為 Arduino 針對 PC 端介面所整合出來的產品。本身以 Leonardo 為主要架構，周邊加上各類型感測器如：聲音、光線、雙軸 PS2 搖桿、按鈕..等，相當適合與 PC 端結合的快速開發。

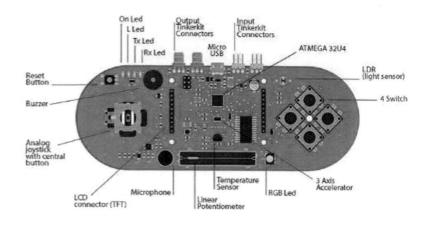

圖 13 Arduino Esplora 控制器

Arduino Esplora 可是為 Arduino 針對 PC 端介面所整合出來的產品，其控制器上包含下列組件：

- 雙軸類比搖桿+按壓開關
- 4 組按鈕開關，以搖桿按鈕的排序呈現
- 線性滑動電阻
- 麥克風聲音感測器
- 光線感測器
- 溫度感測器
- 三軸加速度計
- 蜂鳴器
- RGB LED 燈
- 2 組類比式感測器 輸入擴充腳位
- 2 組數位式輸出擴充腳位
- TFT 顯示螢幕插槽(不含 TFT 螢幕)，可搭配 TFT 螢幕模組使用
- SD 卡擴充插槽(不含 SD 卡相關電路，得透過 TFT 螢幕模組使用)

系統規格

- 核心晶片 - ATmega32U4
- 操作電壓 - 5V
- 輸入電壓 - USB 供電 +5V

- 數位腳位 I/O Pins - 僅存 2 組輸入、2 組輸出可外部擴充
- 類比腳位 - 僅存 2 組輸入可外部擴充
- Flash Memory - 32 KB
- SRAM - 2.5 KB
- EEPROM - 1 KB
- 振盪器頻率 - 16 MHz

圖 14 Arduino Esplora 套件組外觀圖

章節小結

本章節概略的介紹 Arduino 常見的開發板與硬體介紹，接下來就是介紹 Arduino 開發環境，讓我們視目以待。

2

CHAPTER

Arduino 開發環境

Arduino 開發 IDE 安裝

Step1. 進入到 Arduino 官方網站的下載頁面

(http://arduino.cc/en/Main/Software)

Step2. Arduino 的開發環境，有 Windows、Mac OS X、Linux 版本。本範例以 Windows 版本作為範例，請頁面下方點選「Windows Installer」下載 Windows 版本的開發環境。

Arduino IDE

Arduino 1.0.5

Download

Arduino 1.0.5 (release notes), hosted by Google Code:

NOTICE: Arduino Drivers have been updated to add support for Windows 8.1, you can download the updated IDE (version 1.0.5-r2 for Windows) from the download links below.

- Windows Installer, Windows ZIP file (for non-administrator install)
- Mac OS X
- Linux: 32 bit, 64 bit
- source

Next steps

Getting Started

Reference

Environment

Examples

Foundations

FAQ

Step3. 下載完的檔名為「arduino-1.0.5-r2-windows.exe」，將檔案點擊兩下執行，出現如下畫面：

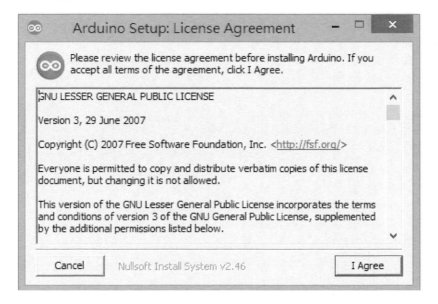

Step4. 點選「I Agree」後出現如下畫面：

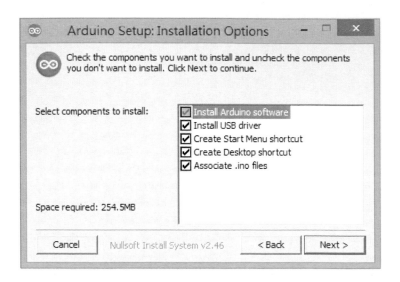

Step5. 點選「Next>」後出現如下畫面：

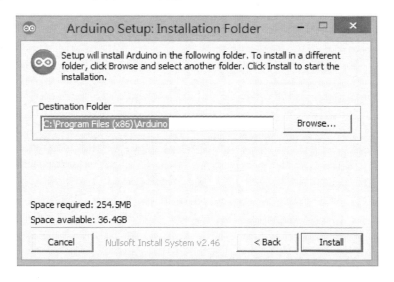

Step6. 選擇檔案儲存位置後，點選「Install」進行安裝，出現如下畫面：

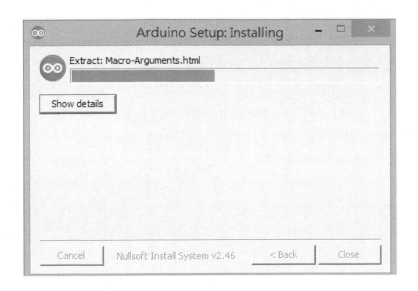

Step7. 安裝到一半時，會出現詢問是否要安裝 Arduino USB Driver(Arduino LLC)的畫面，請點選「安裝(I)」。

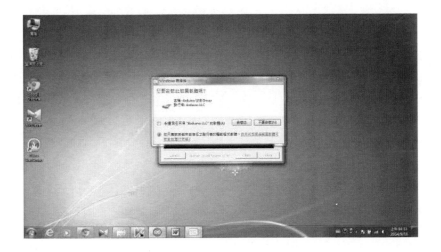

Step8. 安裝系統就會安裝 Arduino USB 驅動程式。

Step9. 安裝完成後，出現如下畫面，點選「Close」。

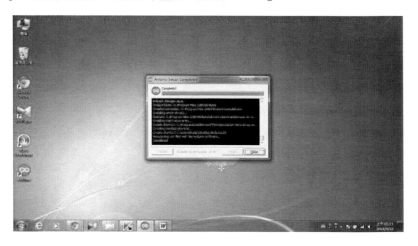

Step10. 桌布上會出現 的圖示，您可以點選該圖示執行 Arduino Sketch 城式。

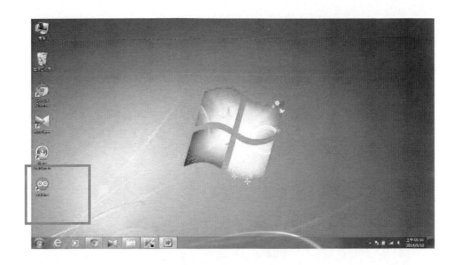

Step11. 您會進入到 Arduino 的軟體開發環境的介面。

以下介紹工具列下方各按鈕的功能：

	Verify 按鈕	進行編譯，驗證程式是否正常運作。
	Upload 按鈕	進行上傳，從電腦把程式上傳到 Arduino 板子裡。
	New 按鈕	新增檔案
	Open 按鈕	開啟檔案，可開啟內建的程式檔或其他檔案。
	Save 按鈕	儲存檔案

Step12. 首先，您可以切換 Arduino Sketch 介面語言。

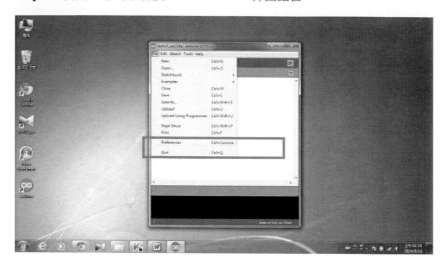

Step13. 出現 Preference 選項畫面。

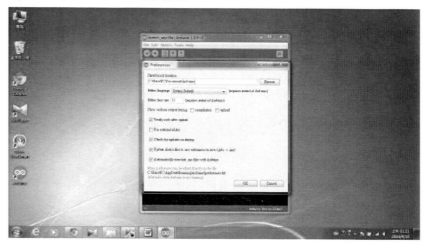

Step14. 可切換到您想要的介面語言(如繁體中文)。

Step15. 切換繁體中文介面語言，按下「OK」。

Step16. 按下「結束鍵」，結束 Arduino Sketch 程式，並重新開啟 Arduino Sketch 程式。

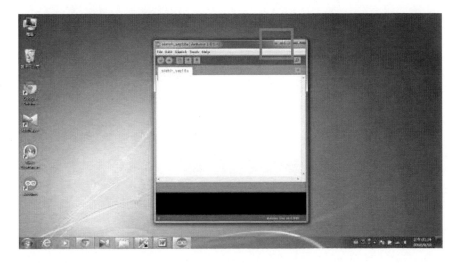

Step17. 可以發現 Arduino Sketch 程式介面語言已經變成繁體中文介面了。

Step18. 點選工具列「草稿碼」中的「匯入程式庫」，並點選「Add Library」選項。

安裝 Arduino 開發板的 USB 驅動程式

以 Mega2560 作為範例

Step1. 將 Mega2560 開發板透過 USB 連接線接上電腦。

Step2. 到剛剛解壓縮完後開啟的資料夾中，點選「drivers」資料夾並進入。

名稱	修改日期	類型	大小
drivers	2014/1/8 下午 08...	檔案資料夾	
examples	2014/1/8 下午 08...	檔案資料夾	
hardware	2014/1/8 下午 08...	檔案資料夾	
java	2014/1/8 下午 08...	檔案資料夾	
lib	2014/1/8 下午 08...	檔案資料夾	
libraries	2014/1/8 下午 08...	檔案資料夾	
reference	2014/1/8 下午 08...	檔案資料夾	
tools	2014/1/8 下午 08...	檔案資料夾	
arduino	2014/1/8 下午 08...	應用程式	840 KB
cygiconv-2.dll	2014/1/8 下午 08...	應用程式擴充	947 KB
cygwin1.dll	2014/1/8 下午 08...	應用程式擴充	1,829 KB
libusb0.dll	2014/1/8 下午 08...	應用程式擴充	43 KB
revisions	2014/1/8 下午 08...	文字文件	38 KB
rxtxSerial.dll	2014/1/8 下午 08...	應用程式擴充	76 KB

Step3. 依照不同位元的作業系統，進行開發板的 USB 驅動程式的安裝。32 位元的作業系統使用 dPinst-x86.exe，64 位元的作業系統使用 dPinst-amd64.exe。

名稱	修改日期	類型	大小
FTDI USB Drivers	2014/1/8 下午 08...	檔案資料夾	
arduino	2014/1/8 下午 08...	安全性目錄	10 KB
arduino	2014/1/8 下午 08...	安裝資訊	7 KB
dpinst-amd64	2014/1/8 下午 08...	應用程式	1,024 KB
dpinst-x86	2014/1/8 下午 08...	應用程式	901 KB
Old Arduino Drivers	2014/1/8 下午 08...	WinRAR ZIP 壓縮檔	14 KB
README	2014/1/8 下午 08...	文字文件	1 KB

Step4. 以 64 位元的作業系統作為範例，點選 dPinst-amd64.exe，會出現如下

畫面：

Step5. 點選「下一步」，程式會進行安裝。完成後出現如下畫面，並點選「完成」。

Step6. 您可至 Arduino 開發環境中工具列「工具」中的「序列埠」看到多出一個 COM，即完成開發板的 USB 驅動程式的設定。

或可至電腦的裝置管理員中，看到連接埠中出現 Arduino Mega 2560 的 COM3，即完成開發板的 USB 驅動程式的設定。

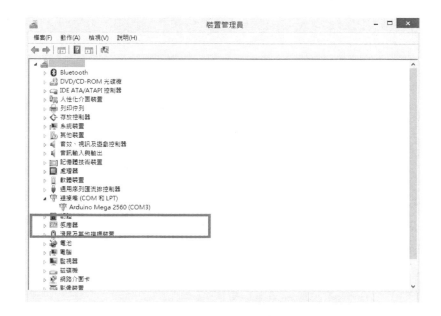

Step7. 到工具列「工具」中的「板子」設定您所用的開發板。

※您可連接多塊 Arduino 開發板至電腦，但工具列中「板子」中的 Board 需與「序列埠」對應。

修改 IDE 開發環境個人喜好設定：(存檔路徑、語言、字型)

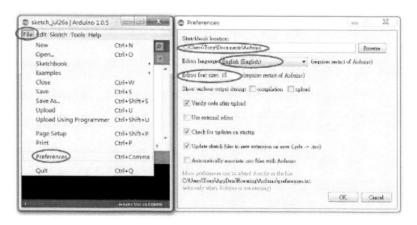

圖 15 IDE 開發環境個人喜好設定

Arduino 函式庫安裝

　　本書使用的 Arduino 函式庫安裝文件，乃以 adafruit 公司官網資料
(https://github.com/adafruit)，的函式庫為範例，進行安裝，展示給各位讀者，首先
讀者可以在 Google 搜尋關鍵字『adafruit　lib』，可以搜尋到 adafruit 公司 Github
網址：https://github.com/adafruit，請讀者選任何一個函示庫。

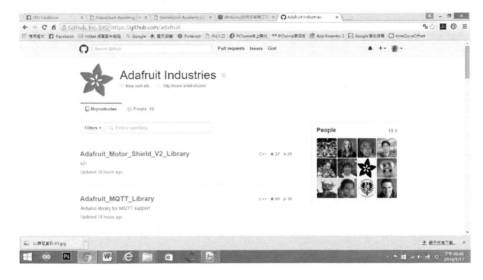

圖　16 Adafruit_Github 官網

　　如下圖所示，本書使用 MQTT 函式庫來當範例，請點選 MQTT 函式庫。

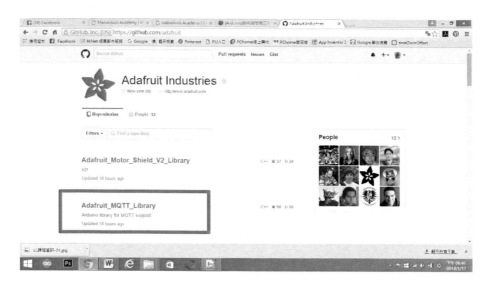

圖 17 點選 MQTT 函式庫

如下圖所示，我們進到 MQTT 函式庫的內容畫面。

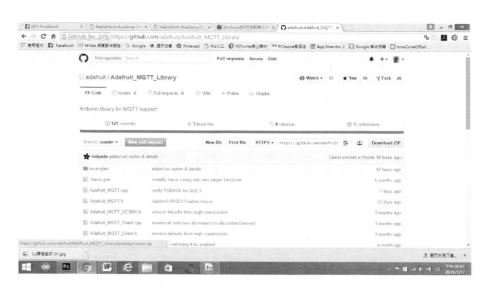

圖 18 MQTT 函數庫畫面

如下圖所示，我們選擇下載 MQTT 函式庫，請點選右上角的『Download Zip』

選項，進行下載函式庫。

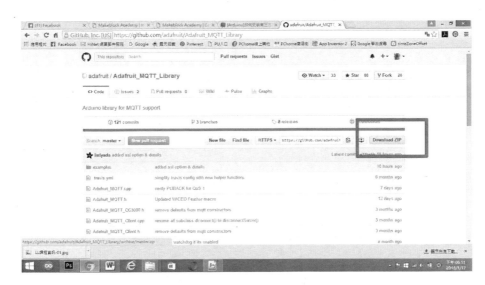

圖 19 點選下載 MQTT 函式庫

如下圖所示，我們下載 MQTT 函式庫完成後，一般而言，都會在系統的下載目錄區內。

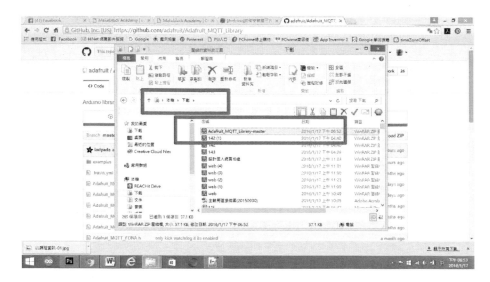

圖 20 MQTT 函數庫壓縮檔下載目錄

如下圖所示，我們進到 Arduino 開發版的開發工具：Sketch IDE 整合環境中。

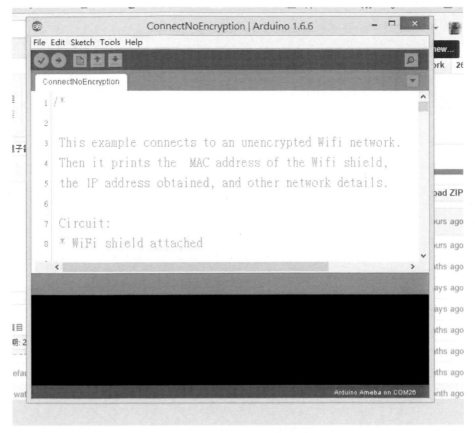

圖 21 Arduino 開發環境

如下圖所示，我們選擇下載加入新的壓縮檔型的函式庫。

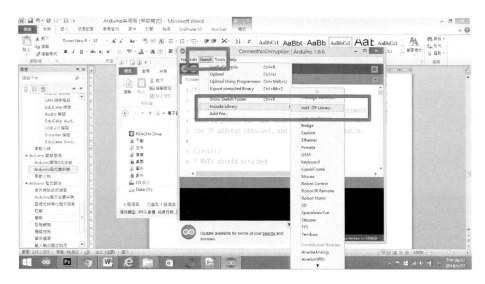

圖 22 加入新函式庫

如下圖所示，我們選擇剛才下載函式庫壓縮檔的目錄，本範例為系統下載目錄。

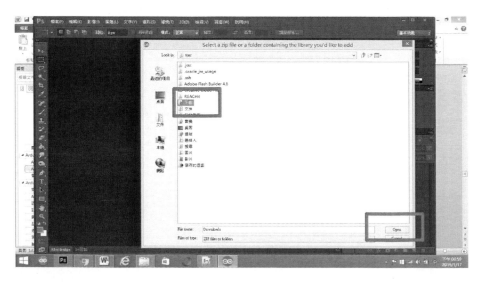

圖 23 選擇函式庫壓縮檔目錄

如下圖所示，我們選擇剛才下載函式庫壓縮檔的目錄，本範例為系統下載目錄

後，我們可以看到該才下載的函式庫壓縮檔，本範例為

『Adafruit_MQTT_Library-master』，請點選 Adafruit_MQTT_Library-master。

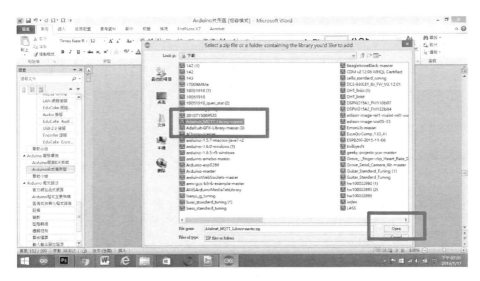

圖 24 選擇函式庫壓縮檔

如下圖所示，我們點選 Adafruit_MQTT_Library-master 之後，回到 Arduino 開

發版的開發工具：Sketch IDE 整合環境中。

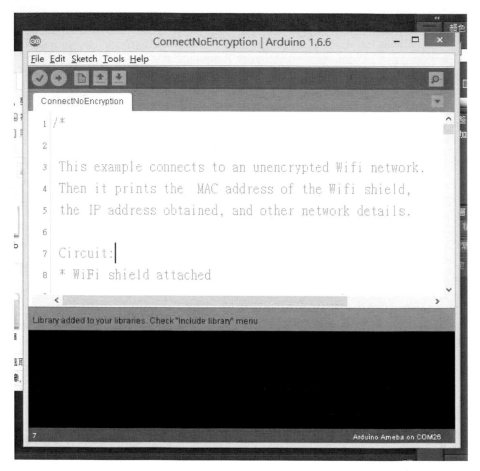

圖 25 安裝好 MQTT 函式庫

　　如下圖所示，進行安裝 Adafruit_MQTT_Library 是否成功安裝，我們使用安裝範例的方法測試，如下圖所示，我們點選 Example→Adafruit_MQTT_Library→mqtt_ethernet 範例程式。

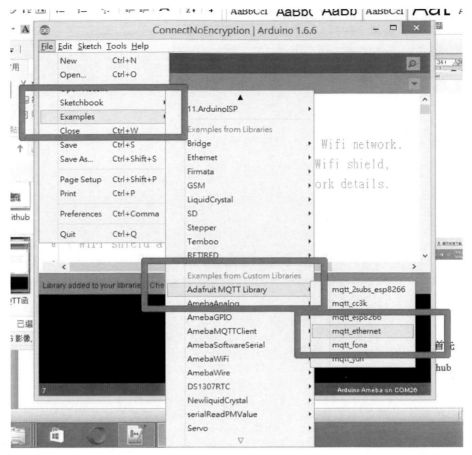

圖 26 測試 MQTT 函式範例程式

　　如下圖所示，如果我們可以正常使用 mqtt_ethernet 範例程式，代表我們已經

將 Adafruit_MQTT_Library 函示庫正確安裝。

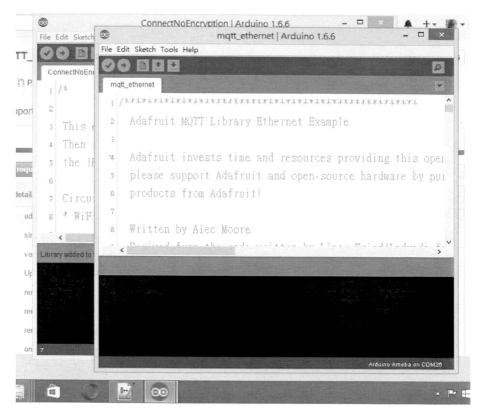

圖 27 安裝 MQTT 函式庫與使用

章節小結

本章節概略的介紹 Arduino 開發環境，主要是讓讀者了解 Arduino 如何操作與撰寫程式。

3

CHAPTER

Arduino 程式語法

官方網站函式網頁

　　讀者若對本章節程式結構不了解之處，請參閱如圖 28 所示之 Arduino 官方網站的 Language Reference (http://arduino.cc/en/Reference/HomePage)，或參閱相關書籍(Anderson & Cervo, 2013; Boxall, 2013; Faludi, 2010; Margolis, 2011, 2012; McRoberts, 2010; Minns, 2013; Monk, 2010, 2012; Oxer & Blemings, 2009; Warren, Adams, & Molle, 2011; Wilcher, 2012)，相信會對 Arduino 程式碼更加了解與熟悉。

Language Reference

Arduino programs can be divided in three main parts: *structure*, *values* (variables and constants), and *functions*.

Structure

+ setup()
+ loop()

Control Structures

+ if
+ if...else
+ for
+ switch case
+ while
+ do... while
+ break
+ continue
+ return
+ goto

Further Syntax

+ ; (semicolon)
+ {} (curly braces)
+ // (single line comment)
+ /* */ (multi-line comment)
+ #define
+ #include

Arithmetic Operators

+ = (assignment operator)
+ + (addition)
+ - (subtraction)
+ * (multiplication)

Variables

Constants

+ HIGH | LOW
+ INPUT | OUTPUT|
INPUT_PULLUP
+ true | false
+ integer constants
+ floating point constants

Data Types

+ void
+ boolean
+ char
+ unsigned char
+ byte
+ int
+ unsigned int
+ word
+ long
+ unsigned long
+ short
+ float
+ double
+ string - char array
+ String - object
+ array

Conversion

+ char()

Functions

Digital I/O

+ pinMode()
+ digitalWrite()
+ digitalRead()

Analog I/O

+ analogReference()
+ analogRead()
+ analogWrite() - PWM

Due only

+ analogReadResolution()
+ analogWriteResolution()

Advanced I/O

+ tone()
+ noTone()
+ shiftOut()
+ shiftIn()
+ pulseIn()

Time

+ millis()
+ micros()
+ delay()
+ delayMicroseconds()

Math

圖 28 Arduino 官方網站的 Language Reference

資料來源：Language Reference (http://arduino.cc/en/Reference/HomePage)

Arduino 程式主要架構

程式結構

> setup()
> loop()

一個 Arduino 程式碼(Sketch)由兩部分組成

setup()

程式初始化

void setup()

在這個函式範圍內放置初始化 Arduino 開發板的程式 - 在重複執行的程式 (loop())之前執行,主要功能是將所有 Arduino 開發板的 Pin 腳設定,元件設定,需要初始化的部分設定等等。

變數型態宣告區 ; // 這裡定義變數或 IO 腳位名稱

void setup()

{

僅在 Power On 或 Reset 後執行一次,setup()函數內放置初始化 Arduino 控制板的程式,即主程式開始執行前需事先設定好的變數 or 腳位定義等例如:PinMode(ledPin,OUTPUT);

}

loop()

迴圈重複執行

void loop()

在此放置你的 Arduino 程式碼。這部份的程式會一直重複的被執行，直到 Arduino 開發板被關閉。

void loop()

```
{


}
```

在 setup()函數之後，即初始化之後，系統則在 loop()程式迴圈內重複執行。直到 Arduino 控制板被關閉。

; 　　　　　　　　**每行程式敘述(statement)後需以分號(";")結束**

{ }(大括號) 　　　**函數前後需用大括號括起來，也可用此將程式碼分成較易讀的區塊**

區塊式結構化程式語言

C 語言是區塊式結構的程式語言， 所謂的區塊是一對大括號：『{}』所界定的範圍， 每一對大括號及其涵括的所有敘述構成 C 語法中所謂的複合敘述 (Compound Statement)， 這樣子的複合敘述不但對於編譯器而言，構成一個有意義的文法單位， 對於程式設計者而言，一個區塊也應該要代表一個完整的程式邏輯單元， 內含的敘述應該具有相當的資料耦合性 (一個敘述處理過的資料會被後面的敘述拿來使用)， 及控制耦合性 (CPU 處理完一個敘述後會接續處理另一個敘述指定的動作)， 當看到程式中一個區塊時， 應該要可以假設其內所包含的敘述都是屬於某些相關功能的， 當然其內部所使用的資料應該都是完成該種功能所必需的， 這些資料應該是專屬於這個區塊內的敘述， 是這個區塊之外的敘述不需要的。

命名空間 (naming space)

C 語言中區塊定義了一塊所謂的命名空間 (naming space)， 在每一個命名空間內，程式設計者可以對其內定義的變數任意取名字， 稱為區域變數 (local variable)， 這些變數只有在該命名空間 (區塊) 內部可以進行存取， 到了該區塊之外程式就不能在藉由該名稱來存取了， 如下例中 int 型態的變數 z。 由於區塊是階層式的， 大區塊可以內含小區塊， 大區塊內的變數也可以在內含區塊內使用， 例如：

```
{
    int x, r;
    x=10;
    r=20;
    {
        int y, z;
        float r;
        y = x;
        x = 1;
        r = 10.5;
    }
    z = x; // 錯誤，不可使用變數 z
}
```

上面這個例子裡有兩個區塊， 也就有兩個命名空間， 有任一個命名空間中不可有兩個變數使用相同的名字， 不同的命名空間則可以取相同的名字， 例如變數 r， 因此針對某一個變數來說， 可以使用到這個變數的程式範圍就稱為這個變數的作用範圍 (scope)。

變數的生命期 (Lifetime)

變數的生命始於定義之敘述而一直延續到定義該變數之區塊結束為止，變數的作用範圍：意指程式在何處可以存取該變數，有時變數是存在的，但是程式卻無法藉由其名稱來存取它，例如，上例中內層區塊內無法存取外層區塊所定義的變數 r，因為在內層區塊中 r 這個名稱賦予另一個 float 型態的變數了。

縮小變數的作用範圍

利用 C 語言的區塊命名空間的設計，程式設計者可以儘量把變數的作用範圍縮小，如下例：

```
{
int tmp;
    for (tmp=0; tmp<1000; tmp++)
        doSomeThing();
}
{
    float tmp;
    tmp = y;
    y = x;
    x = y;
}
```

上面這個範例中前後兩個區塊中的 tmp 很明顯地沒有任何關係，看這個程式的人不必擔心程式中有藉 tmp 變數傳遞資訊的任何意圖。

特殊符號

; (semicolon)
{} (curly braces)
// (single line comment)
/* */ (multi-line comment)

Arduino 語言用了一些符號描繪程式碼，例如註解和程式區塊。

; //(分號)

Arduino 語言每一行程序都是以分號為結尾。這樣的語法讓你可以自由地安排代碼，你可以將兩個指令放置在同一行，只要中間用分號隔開（但這樣做可能降低程式的可讀性）。

範例：

```
delay(100);
```

{}(大括號)

大括號用來將程式代碼分成一個又一個的區塊，如以下範例所示，在 loop() 函式的前、後，必須用大括號括起來。

範例：

```
void loop(){
    Serial.pritln("Hello !! Welcome to Arduino world");
}
```

註解

程式的註解就是對代碼的解釋和說明，攥寫註解有助於程式設計師(或其他人)了解代碼的功能。

Arduino 處理器在對程式碼進行編譯時會忽略註解的部份。

Arduino 語言中的攥寫註解有兩種方式

```
//單行註解：這整行的文字會被處理器忽略
/*多行註解：
    在這個範圍內你可以
    寫  一篇  小說
```

```
*/
```

變數

　　程式中的變數與數學使用的變數相似，都是用某些符號或單字代替某些數值，從而得以方便計算過程。程式語言中的變數屬於識別字 (identifier) ，C 語言對於識別字有一定的命名規則，例如只能用英文大小寫字母、數字以及底線符號

　　其中，數字不能用作識別字的開頭，單一識別字裡不允許有空格，而如 int 、char 為 C 語言的關鍵字 (keyword) 之一，屬於程式語言的語法保留字，因此也不能用為自行定義的名稱。通常編譯器至少能讀取名稱的前 31 個字元，但外部名稱可能只能保證前六個字元有效。

　　變數使用前要先進行宣告 (declaration) ，宣告的主要目的是告訴編譯器這個變數屬於哪一種資料型態，好讓編譯器預先替該變數保留足夠的記憶體空間。宣告的方式很簡單，就是型態名稱後面接空格，然後是變數的識別名稱

常數

➢ HIGH | LOW
➢ INPUT | OUTPUT
➢ true | false
➢ Integer Constants

資料型態

➢ boolean
➢ char
➢ byte
➢ int
➢ unsigned int
➢ long
➢ unsigned long
➢ float

> ➢ double
> ➢ string
> ➢ array
> ➢ void

常數

在 Arduino 語言中事先定義了一些具特殊用途的保留字。HIGH 和 LOW 用來表示你開啟或是關閉了一個 Arduino 的腳位(Pin)。INPUT 和 OUTPUT 用來指示這個 Arduino 的腳位(Pin)是屬於輸入或是輸出用途。true 和 false 用來指示一個條件或表示式為真或是假。

變數

變數用來指定 Arduino 記憶體中的一個位置,變數可以用來儲存資料,程式人員可以透過程式碼去不限次數的操作變數的值。

因為 Arduino 是一個非常簡易的微處理器,但你要宣告一個變數時必須先定義他的資料型態,好讓微處理器知道準備多大的空間以儲存這個變數值。

Arduino 語言支援的資料型態:

布林 boolean

布林變數的值只能為真(true)或是假(false)

字元 char

單一字元例如 A,和一般的電腦做法一樣 Arduino 將字元儲存成一個數字,即使你看到的明明就是一個文字。

用數字表示一個字元時,它的值有效範圍為 -128 到 127。

PS:目前有兩種主流的電腦編碼系統 ASCII 和 UNICODE。

- ASCII 表示了 127 個字元， 用來在序列終端機和分時計算機之間傳輸文字。

- UNICODE 可表示的字量比較多，在現代電腦作業系統內它可以用來表示多國語言。

在位元數需求較少的資訊傳輸時，例如義大利文或英文這類由拉丁文，阿拉伯數字和一般常見符號構成的語言，ASCII 仍是目前主要用來交換資訊的編碼法。

位元組 byte

儲存的數值範圍為 0 到 255。如同字元一樣位元組型態的變數只需要用一個位元組(8 位元)的記憶體空間儲存。

整數 int

整數資料型態用到 2 位元組的記憶體空間，可表示的整數範圍為 −32,768 到 32,767; 整數變數是 Arduino 內最常用到的資料型態。

整數 unsigned int

無號整數同樣利用 2 位元組的記憶體空間，無號意謂著它不能儲存負的數值，因此無號整數可表示的整數範圍為 0 到 65,535。

長整數 long

長整數利用到的記憶體大小是整數的兩倍，因此它可表示的整數範圍從 −2,147,483,648 到 2,147,483,647。

長整數 unsigned long

無號長整數可表示的整數範圍為 0 到 4,294,967,295。

浮點數 float

浮點數就是用來表達有小數點的數值，每個浮點數會用掉四位元組的 RAM，注意晶片記憶體空間的限制，謹慎的使用浮點數。

雙精準度 浮點數 double

雙精度浮點數可表達最大值為 1.7976931348623157 x 10308。

字串 string

字串用來表達文字信息，它是由多個 ASCII 字元組成(你可以透過序串埠發送一個文字資訊或者將之顯示在液晶顯示器上)。字串中的每一個字元都用一個組元組空間儲存，並且在字串的最尾端加上一個空字元以提示 Ardunio 處理器字串的結束。下面兩種宣告方式是相同的。

```
char word1 = "Arduino world"; // 7 字元 ＋1 空字元
char word2 = "Arduino is a good developed kit"; // 與上行相同
```

陣列 array

一串變數可以透過索引去直接取得。假如你想要儲存不同程度的 LED 亮度時，你可以宣告六個變數 light01，light02，light03，light04，light05，light06，但其實你有更好的選擇，例如宣告一個整數陣列變數如下：

```
int light = {0, 20, 40, 65, 80, 100};
```

"array" 這個字為沒有直接用在變數宣告，而是[]和{}宣告陣列。

控制指令

string(字串)

範例

```
char Str1[15];
```

```
char Str2[8] = {'a', 'r', 'd', 'u', 'i', 'n', 'o'};
char Str3[8] = {'a', 'r', 'd', 'u', 'i', 'n', 'o', '\0'};
char Str4[ ] = "arduino";
char Str5[8] = "arduino";
char Str6[15] = "arduino";
```

解釋如下:

- 在 Str1 中 聲明一個沒有初始化的字元陣列

- 在 Str2 中 聲明一個字元陣列(包括一個附加字元),編譯器會自動添加所需的空字元

- 在 Str3 中 明確加入空字元

- 在 Str4 中 用引號分隔初始化的字串常數,編譯器將調整陣列的大小,以適應字串常量和終止空字元

- 在 Str5 中 初始化一個包括明確的尺寸和字串常量的陣列

- 在 Str6 中 初始化陣列,預留額外的空間用於一個較大的字串

空終止字元

一般來說,字串的結尾有一個空終止字元(ASCII 代碼 0), 以此讓功能函數(例如 Serial.prinf())知道一個字串的結束, 否則,他們將從記憶體繼續讀取後續位元組,而這些並不屬於所需字串的一部分。

這表示你的字串比你想要的文字包含更多的個字元空間, 這就是為什麼 Str2 和 Str5 需要八個字元, 即使"Arduino"只有七個字元 - 最後一個位置會自動填充空字元, str4 將自動調整為八個字元,包括一個額外的 null, 在 Str3 的,我們自己已經明確地包含了空字元(寫入'\0')。

使用符號:單引號?還是雙引號?

- 定義字串時使用雙引號(例如"ABC")，

- 定義一個單獨的字元時使用單引號(例如'A')

範例

<table>
<tr><td>字串測試範例(stringtest01)</td></tr>
<tr><td>

```
char* myStrings[]={
  "This is string 1", "This is string 2", "This is string 3",
  "This is string 4", "This is string 5","This is string 6"};

void setup(){
  Serial.begin(9600);
}

void loop(){
  for (int i = 0; i < 6; i++){
    Serial.println(myStrings[i]);
    delay(500);
  }
}
```

</td></tr>
</table>

char* 在字元資料類型 char 後跟了一個星號'*'表示這是一個"指標"陣列，所有的陣列名稱實際上是指標，所以這需要一個陣列的陣列。

指標對於 C 語言初學者而言是非常深奧的部分之一，但是目前我們沒有必要瞭解詳細指標，就可以有效地應用它。

型態轉換

- ➢ char()
- ➢ byte()
- ➢ int()
- ➢ long()
- ➢ float()

char()

指令用法

將資料轉程字元形態：

語法：char(x)

參數

x: 想要轉換資料的變數或內容

回傳

字元形態資料

unsigned char()

一個無符號資料類型佔用 1 個位元組的記憶體:與 byte 的資料類型相同，無符號的 char 資料類型能編碼 0 到 255 的數位，為了保持 Arduino 的程式設計風格的一致性，byte 資料類型是首選。

指令用法

將資料轉程字元形態：

語法：unsigned char(x)

參數

x: 想要轉換資料的變數或內容

回傳

字元形態資料

```
unsigned char myChar = 240;
```

byte()

指令用法

將資料轉換位元資料形態：

語法：bytc(x)

參數

x: 想要轉換資料的變數或內容

回傳

位元資料形態的資料

int(x)

指令用法

將資料轉換整數資料形態：

語法：int(x)

參數

x: 想要轉換資料的變數或內容

回傳

整數資料形態的資料

unsigned int(x)

unsigned int(無符號整數)與整型資料同樣大小，佔據 2 位元組: 它只能用於存儲正數而不能存儲負數，範圍 0~65,535 (2^16) - 1)。

指令用法

將資料轉換整數資料形態：

語法：unsigned int(x)

參數

x: 想要轉換資料的變數或內容

回傳

整數資料形態的資料

```
unsigned int ledPin = 13;
```

long()

指令用法

將資料轉換長整數資料形態：

語法：int(x)

參數

x: 想要轉換資料的變數或內容

回傳

長整數資料形態的資料

unsigned long()

無符號長整型變數擴充了變數容量以存儲更大的資料， 它能存儲 32 位元(4 位元組)資料:與標準長整型不同無符號長整型無法存儲負數， 其範圍從 0 到 4,294,967,295（$2^{32}-1$）。

指令用法

將資料轉換長整數資料形態：

語法：unsigned int(x)

參數

x: 想要轉換資料的變數或內容

回傳

長整數資料形態的資料

```
unsigned long time;

void setup()
{
    Serial.begin(9600);
```

```
}

void loop()
{
    Serial.print("Time: ");
    time = millis();
    //程式開始後一直列印時間
    Serial.println(time);
    //等待一秒鐘，以免發送大量的資料
    delay(1000);
}
```

float()

指令用法

將資料轉換浮點數資料形態：

語法：float(x)

參數

x: 想要轉換資料的變數或內容

回傳

浮點數資料形態的資料

邏輯控制

控制流程

if
if...else
for
switch case
while
do... while
break
continue

return

Ardunio 利用一些關鍵字控制程式碼的邏輯。

if … else

If 必須緊接著一個問題表示式(expression)，若這個表示式為真，緊連著表示式後的代碼就會被執行。若這個表示式為假，則執行緊接著 else 之後的代碼. 只使用 if 不搭配 else 是被允許的。

範例：

```
#define LED 12
void setup()
{
    int val =1;
    if (val == 1) {
    digitalWrite(LED,HIGH);
}
}
void loop()
{
}
```

for

用來明定一段區域代碼重覆指行的次數。

範例：

```
void setup()
{
    for (int i = 1; i < 9; i++) {
        Serial.print("2 * ");
        Serial.print(i);
        Serial.print(" = ");
        Serial.print(2*i);
```

```
    }
}
void loop()
{
}
```

switch case

if 敘述是程式裡的分叉選擇，switch case 是更多選項的分叉選擇。swith case 根據變數值讓程式有更多的選擇，比起一串冗長的 if 敘述，使用 swith case 可使程式代碼看起來比較簡潔。

範例：

```
void setup()
{
    int sensorValue;
    sensorValue = analogRead(1);
    switch (sensorValue) {

    case 10:
      digitalWrite(13,HIGH);
      break;

case 20:
    digitalWrite(12,HIGH);
    break;

default: // 以上條件都不符合時，預設執行的動作
    digitalWrite(12,LOW);
    digitalWrite(13,LOW);
}
}
void loop()
{

    }
```

while

當 while 之後的條件成立時，執行括號內的程式碼。

範例：

```
void setup()
{
  int sensorValue;
  // 當 sensor 值小於 256，閃爍 LED 1 燈
  sensorValue = analogRead(1);
  while (sensorValue < 256) {
    digitalWrite(13,HIGH);
    delay(100);
    digitalWrite(13,HIGH);
    delay(100);
    sensorValue = analogRead(1);
  }
}
void loop()
{
  }
```

do … while

和 while 相似，不同的是 while 前的那段程式碼會先被執行一次，不管特定的條件式為真或為假。因此若有一段程式代碼至少需要被執行一次，就可以使用 do…while 架構。

範例：

```
void setup()
{
  int sensorValue;
  do
  {
    digitalWrite(13,HIGH);
    delay(100);
    digitalWrite(13,HIGH);
```

```
    delay(100);
    sensorValue = analogRead(1);
  }
  while (sensorValue < 256);
}
void loop()
{
}
```

break

Break 讓程式碼跳離迴圈，並繼續執行這個迴圈之後的程式碼。此外，在 break 也用於分隔 switch case 不同的敘述。

範例：

```
void setup()
{
}
void loop()
{
  int sensorValue;
  do {
    // 按下按鈕離開迴圈
    if (digitalRead(7) == HIGH)
        break;
        digitalWrite(13,HIGH);
        delay(100);
        digitalWrite(13,HIGH);
        delay(100);
        sensorValue = analogRead(1);
  }
  while (sensorValue < 512);
}
```

continue

continue 用於迴圈之內，它可以強制跳離接下來的程式，並直接執行下一個迴圈。

範例：

```
#define PWMPin 12
#define SensorPin 8
void setup()
{
}
void loop()
{
  int light;
  int x ;
  for (light = 0; light < 255; light++)
  {
     // 忽略數值介於 140 到 200 之間
      x = analogRead(SensorPin) ;

    if ((x > 140) && (x < 200))
      continue;

    analogWrite(PWMPin, light);
    delay(10);

  }
}
```

return

函式的結尾可以透過 return 回傳一個數值。

例如，有一個計算現在溫度的函式叫 computeTemperature()，你想要回傳現在的溫度給 temperature 變數，你可以這樣寫：

```
#define PWMPin 12
#define SensorPin 8

void setup()
{
}
void loop()
```

```
{
  int light;
  int x ;
  for (light = 0; light < 255; light++)
  {
    // 忽略數值介於 140 到 200 之間
    x = computeTemperature() ;
    if ((x > 140) && (x < 200))
        continue;

        analogWrite(PWMPin, light);
        delay(10);
  }
}
int computeTemperature() {

  int temperature = 0;
  temperature = (analogRead(SensorPin) + 45) / 100;
      return temperature;
}
```

算術運算

算術符號

$=$　(給值)

$+$　(加法)

$-$　(減法)

$*$　(乘法)

$/$　(除法)

$\%$　(求餘數)

你可以透過特殊的語法用 Arduino 去做一些複雜的計算。 $+$ 和 $-$ 就是一般數學上的加減法，乘法用*示，而除法用 /表示。

另外餘數除法(%)，用於計算整數除法的餘數值: 一個整數除以另一個數，其餘數稱為模數，它有助於保持一個變數在一個特定的範圍(例如陣列的大小)。

語法：

result = dividend % divisor

參數：

● dividend：一個被除的數字
● divisor：一個數字用於除以其他數

{}括號

你可以透過多層次的括弧去指定算術之間的循序。和數學函式不一樣，中括號和大括號在此被保留在不同的用途(分別為陣列索引，和宣告區域程式碼)。

範例：

```
#define PWMPin 12
#define SensorPin 8

void setup()
{
     int sensorValue;
     int light;
     int remainder;

     sensorValue = analogRead(SensorPin) ;
     light = ((12 * sensorValue) - 5 ) / 2;
     remainder = 3 % 2;

}
void loop()
{
}
```

比較運算

== (等於)

!= (不等於)

< (小於)

> (大於)

<= (小於等於)

>= (大於等於)

當你在指定 if,while, for 敘述句時，可以運用下面這個運算符號：

符號	意義	範例
==	等於	a==1
!=	不等於	a!=1
<	小於	a<1
>	大於	a>1
<=	小於等於	a<=1
>=	大於等於	a>=1

布林運算

➢ && (and)
➢ || (or)
➢ ! (not)

當你想要結合多個條件式時，可以使用布林運算符號。

例如你想要檢查從感測器傳回的數值是否於 5 到 10，你可以這樣寫：

```
#define PWMPin 12
#define SensorPin 8
void setup()
{
}
void loop()
{
  int light;
```

```
int sensor ;
for (light = 0; light < 255; light++)
{
        // 忽略數值介於 140 到 200 之間
         sensor = analogRead(SensorPin) ;

if ((sensor >= 5) && (sensor <=10))
     continue;

     analogWrite(PWMPin, light);
     delay(10);
}
}
```

這裡有三個運算符號: 交集(and)用 **&&** 表示; 聯集(or)用 ‖ 表示; 反相 (finally not)用 !表示。

複合運算符號:有一般特殊的運算符號可以使程式碼比較簡潔,例如累加運算符號。

例如將一個值加 1,你可以這樣寫:

```
Int value = 10 ;
value = value + 1 ;
```

你也可以用一個複合運算符號累加(++):

```
Int value = 10 ;
value ++;
```

複合運算符號

- ➢   ++ (increment)
- ➢   -- (decrement)
- ➢   += (compound addition)
- ➢   -= (compound subtraction)

- ➢ *= (compound multiplication)
- ➢ /= (compound division)

累加和遞減 (++ 和 --)

當你在累加 1 或遞減 1 到一個數值時。請小心 i++ 和 ++i 之間的不同。如果你用的是 i++，i 會被累加並且 i 的值等於 i+1；但當你使用 ++i 時，i 的值等於 i，直到這行指令被執行完時 i 再加 1。同理應用於—。

+= , -=, *= and /=

這些運算符號可讓表示式更精簡，下面二個表示式是等價的：

```
Int value = 10 ;
value   = value +5 ;        // (此兩者都是等價)
value   += 5 ;              // (此兩者都是等價)
```

輸入輸出腳位設定

數位訊號輸出/輸入

- ➢ PinMode()
- ➢ digitalWrite()
- ➢ digitalRead()

類比訊號輸出/輸入

- ➢ analogRead()
- ➢ analogWrite() - PWM

Arduino 內含了一些處理輸出與輸入的切換功能，相信已經從書中程式範例略知一二。

PinMode(Pin, mode)

將數位腳位(digital Pin)指定為輸入或輸出。

範例

```
#define sensorPin 7
#define PWNPin 8
void setup()
{
PinMode(sensorPin,INPUT); // 將腳位 sensorPin (7) 定為輸入模式
}
void loop()
{
}
```

digitalWrite(Pin, value)

將數位腳位指定為開或關。腳位必須先透過 PinMode 明示為輸入或輸出模式 digitalWrite 才能生效。

範例：

```
#define PWNPin 8
#define sensorPin 7
void setup()
{
digitalWrite (PWNPin,OUTPUT); // 將腳位 PWNPin (8) 定為輸入模式
}
void loop()
{}
```

int digitalRead(Pin)

將輸入腳位的值讀出，當感測到腳位處於高電位時時回傳 HIGH，否則回傳 LOW。

範例：

```
#define PWNPin 8
```

```
#define sensorPin 7
void setup()
{
    PinMode(sensorPin,INPUT); // 將腳位 sensorPin (7) 定為輸入模式
    val = digitalRead(7); // 讀出腳位 7 的值並指定給 val
}
void loop()
{
}
```

int analogRead(Pin)

讀出類比腳位的電壓並回傳一個 0 到 1023 的數值表示相對應的 0 到 5 的電壓值。

範例：

```
#define PWNPin 8
#define sensorPin 7
void setup()
{
    PinMode(sensorPin,INPUT); // 將腳位 sensorPin (7) 定為輸入模式
    val = analogRead (7); // 讀出腳位 7 的值並指定給 val
}
void loop()
{
}
```

analogWrite(Pin, value)

改變 PWM 腳位的輸出電壓值，腳位通常會在 3、5、6、9、10 與 11。value 變數範圍 0-255，例如：輸出電壓 2.5 伏特（V），該值大約是 128。

範例：

```
#define PWNPin 8
#define sensorPin 7
void setup()
```

```
{
analogWrite (PWNPin,OUTPUT); // 將腳位 PWNPin (8) 定為輸入模式
}
void loop()
{     }
```

進階 I/O

> tone()
> noTone()
> shiftOut()
> pulseIn()

tone(Pin)

使用 Arduino 開發板，使用一個 Digital Pin(數位接腳)連接喇叭，如本例子是接在數位接腳 13(Digital Pin 13)，讀者也可將喇叭接在您想要的腳位，只要將下列程式作對應修改，可以產生想要的音調。

範例 :

```
#include <Tone.h>

Tone tone1;

void setup()
{
  tone1.begin(13);
  tone1.play(NOTE_A4);
}

void loop()
{
}
```

表 1 Tone 頻率表

常態變數	頻率(Frequency (Hz))
NOTE_B2	123
NOTE_C3	131
NOTE_CS3	139
NOTE_D3	147
NOTE_DS3	156
NOTE_E3	165
NOTE_F3	175
NOTE_FS3	185
NOTE_G3	196
NOTE_GS3	208
NOTE_A3	220
NOTE_AS3	233
NOTE_B3	247
NOTE_C4	262
NOTE_CS4	277
NOTE_D4	294
NOTE_DS4	311
NOTE_E4	330
NOTE_F4	349
NOTE_FS4	370
NOTE_G4	392
NOTE_GS4	415
NOTE_A4	440
NOTE_AS4	466
NOTE_B4	494
NOTE_C5	523
NOTE_CS5	554
NOTE_D5	587
NOTE_DS5	622
NOTE_E5	659
NOTE_F5	698
NOTE_FS5	740
NOTE_G5	784
NOTE_GS5	831
NOTE_A5	880

常態變數	頻率(Frequency (Hz))
NOTE_AS5	932
NOTE_B5	988
NOTE_C6	1047
NOTE_CS6	1109
NOTE_D6	1175
NOTE_DS6	1245
NOTE_E6	1319
NOTE_F6	1397
NOTE_FS6	1480
NOTE_G6	1568
NOTE_GS6	1661
NOTE_A6	1760
NOTE_AS6	1865
NOTE_B6	1976
NOTE_C7	2093
NOTE_CS7	2217
NOTE_D7	2349
NOTE_DS7	2489
NOTE_E7	2637
NOTE_F7	2794
NOTE_FS7	2960
NOTE_G7	3136
NOTE_GS7	3322
NOTE_A7	3520
NOTE_AS7	3729
NOTE_B7	3951
NOTE_C8	4186
NOTE_CS8	4435
NOTE_D8	4699
NOTE_DS8	4978

資料來源：

https://code.google.com/p/rogue-code/wiki/ToneLibraryDocumentation#Ugly_Details

表 2 Tone 音階頻率對照表

音階	常態變數	頻率(Frequency (Hz))
低音 Do	NOTE_C4	262
低音 Re	NOTE_D4	294
低音 Mi	NOTE_E4	330
低音 Fa	NOTE_F4	349
低音 So	NOTE_G4	392
低音 La	NOTE_A4	440
低音 Si	NOTE_B4	494
中音 Do	NOTE_C5	523
中音 Re	NOTE_D5	587
中音 Mi	NOTE_E5	659
中音 Fa	NOTE_F5	698
中音 So	NOTE_G5	784
中音 La	NOTE_A5	880
中音 Si	NOTE_B5	988
高音 Do	NOTE_C6	1047
高音 Re	NOTE_D6	1175
高音 Mi	NOTE_E6	1319
高音 Fa	NOTE_F6	1397
高音 So	NOTE_G6	1568
高音 La	NOTE_A6	1760
高音 Si	NOTE_B6	1976
高高音 Do	NOTE_C7	2093

資料來源：

https://code.google.com/p/rogue-code/wiki/ToneLibraryDocumentation#Ugly_Details

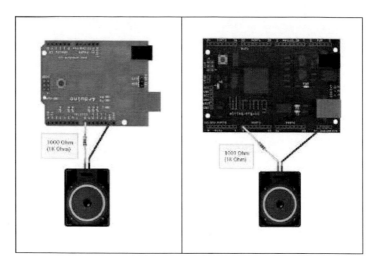

圖 29 Tone 接腳圖

資料來源：

https://code.google.com/p/rogue-code/wiki/ToneLibraryDocumentation#Ugly_Details

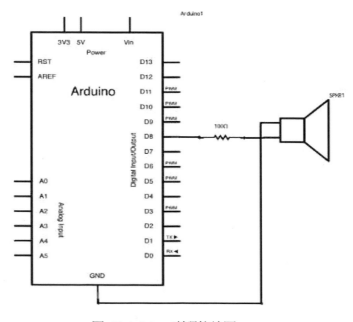

圖 30 Arduino 喇叭接線圖

Mario 音樂範例：

```
/*
   Arduino Mario Bros Tunes
   With Piezo Buzzer and PWM
   by: Dipto Pratyaksa
   last updated: 31/3/13
*/
#include <pitches.h>

#define melodyPin 3
//Mario main theme melody
int melody[] = {
   NOTE_E7, NOTE_E7, 0, NOTE_E7,
   0, NOTE_C7, NOTE_E7, 0,
   NOTE_G7, 0, 0,   0,
   NOTE_G6, 0, 0, 0,

   NOTE_C7, 0, 0, NOTE_G6,
   0, 0, NOTE_E6, 0,
   0, NOTE_A6, 0, NOTE_B6,
   0, NOTE_AS6, NOTE_A6, 0,

   NOTE_G6, NOTE_E7, NOTE_G7,
   NOTE_A7, 0, NOTE_F7, NOTE_G7,
   0, NOTE_E7, 0,NOTE_C7,
   NOTE_D7, NOTE_B6, 0, 0,

   NOTE_C7, 0, 0, NOTE_G6,
   0, 0, NOTE_E6, 0,
   0, NOTE_A6, 0, NOTE_B6,
   0, NOTE_AS6, NOTE_A6, 0,

   NOTE_G6, NOTE_E7, NOTE_G7,
   NOTE_A7, 0, NOTE_F7, NOTE_G7,
   0, NOTE_E7, 0,NOTE_C7,
   NOTE_D7, NOTE_B6, 0, 0
};
//Mario main them tempo
int tempo[] = {
   12, 12, 12, 12,
```

```
    12, 12, 12, 12,
    12, 12, 12, 12,
    12, 12, 12, 12,

    12, 12, 12, 12,
    12, 12, 12, 12,
    12, 12, 12, 12,
    12, 12, 12, 12,

    9, 9, 9,
    12, 12, 12, 12,
    12, 12, 12, 12,
    12, 12, 12, 12,

    12, 12, 12, 12,
    12, 12, 12, 12,
    12, 12, 12, 12,
    12, 12, 12, 12,

    9, 9, 9,
    12, 12, 12, 12,
    12, 12, 12, 12,
    12, 12, 12, 12,
};

//

//Underworld melody
int underworld_melody[] = {
    NOTE_C4, NOTE_C5, NOTE_A3, NOTE_A4,
    NOTE_AS3, NOTE_AS4, 0,
    0,
    NOTE_C4, NOTE_C5, NOTE_A3, NOTE_A4,
    NOTE_AS3, NOTE_AS4, 0,
    0,
    NOTE_F3, NOTE_F4, NOTE_D3, NOTE_D4,
    NOTE_DS3, NOTE_DS4, 0,
    0,
    NOTE_F3, NOTE_F4, NOTE_D3, NOTE_D4,
```

```
    NOTE_DS3, NOTE_DS4, 0,
    0, NOTE_DS4, NOTE_CS4, NOTE_D4,
    NOTE_CS4, NOTE_DS4,
    NOTE_DS4, NOTE_GS3,
    NOTE_G3, NOTE_CS4,
    NOTE_C4, NOTE_FS4,NOTE_F4, NOTE_E3, NOTE_AS4, NOTE_A4,
    NOTE_GS4, NOTE_DS4, NOTE_B3,
    NOTE_AS3, NOTE_A3, NOTE_GS3,
    0, 0, 0
};
//Underwolrd tempo
int underworld_tempo[] = {
    12, 12, 12, 12,
    12, 12, 6,
    3,
    12, 12, 12, 12,
    12, 12, 6,
    3,
    12, 12, 12, 12,
    12, 12, 6,
    3,
    12, 12, 12, 12,
    12, 12, 6,
    6, 18, 18, 18,
    6, 6,
    6, 6,
    6, 6,
    18, 18, 18,18, 18, 18,
    10, 10, 10,
    10, 10, 10,
    3, 3, 3
};

void setup(void)
{
    PinMode(3, OUTPUT);//buzzer
    PinMode(13, OUTPUT);//led indicator when singing a note

}
```

```
void loop()
{
//sing the tunes
  sing(1);
  sing(1);
  sing(2);
}
int song = 0;

void sing(int s){
    // iterate over the notes of the melody:
    song = s;
    if(song==2){
        Serial.println(" 'Underworld Theme'");
        int size = sizeof(underworld_melody) / sizeof(int);
        for (int thisNote = 0; thisNote < size; thisNote++) {

            // to calculate the note duration, take one second
            // divided by the note type.
            //e.g. quarter note = 1000 / 4, eighth note = 1000/8, etc.
            int noteDuration = 1000/underworld_tempo[thisNote];

            buzz(melodyPin, underworld_melody[thisNote],noteDuration);

            // to distinguish the notes, set a minimum time between them.
            // the note's duration + 30% seems to work well:
            int pauseBetweenNotes = noteDuration * 1.30;
            delay(pauseBetweenNotes);

            // stop the tone playing:
            buzz(melodyPin, 0,noteDuration);

        }

    }else{

        Serial.println(" 'Mario Theme'");
        int size = sizeof(melody) / sizeof(int);
        for (int thisNote = 0; thisNote < size; thisNote++) {
```

```
        // to calculate the note duration, take one second
        // divided by the note type.
        //e.g. quarter note = 1000 / 4, eighth note = 1000/8, etc.
        int noteDuration = 1000/tempo[thisNote];

        buzz(melodyPin, melody[thisNote],noteDuration);

        // to distinguish the notes, set a minimum time between them.
        // the note's duration + 30% seems to work well:
        int pauseBetweenNotes = noteDuration * 1.30;
        delay(pauseBetweenNotes);

        // stop the tone playing:
        buzz(melodyPin, 0,noteDuration);

    }
  }
}

void buzz(int targetPin, long frequency, long length) {
  digitalWrite(13,HIGH);
  long delayValue = 1000000/frequency/2; // calculate the delay value between
transitions
  //// 1 second's worth of microseconds, divided by the frequency, then split in half
since
  //// there are two phases to each cycle
  long numCycles = frequency * length/ 1000; // calculate the number of cycles for
proper timing
  //// multiply frequency, which is really cycles per second, by the number of sec-
onds to
  //// get the total number of cycles to produce
  for (long i=0; i < numCycles; i++){ // for the calculated length of time...
    digitalWrite(targetPin,HIGH); // write the buzzer Pin high to push out the dia-
phram
    delayMicroseconds(delayValue); // wait for the calculated delay value
    digitalWrite(targetPin,LOW); // write the buzzer Pin low to pull back the dia-
phram
    delayMicroseconds(delayValue); // wait again or the calculated delay value
```

```
  }
  digitalWrite(13,LOW);


}
```

```
/*************************************************
 * Public Constants
 *************************************************/

#define NOTE_B0    31
#define NOTE_C1    33
#define NOTE_CS1 35
#define NOTE_D1    37
#define NOTE_DS1 39
#define NOTE_E1    41
#define NOTE_F1    44
#define NOTE_FS1 46
#define NOTE_G1    49
#define NOTE_GS1 52
#define NOTE_A1    55
#define NOTE_AS1 58
#define NOTE_B1    62
#define NOTE_C2    65
#define NOTE_CS2 69
#define NOTE_D2    73
#define NOTE_DS2 78
#define NOTE_E2    82
#define NOTE_F2    87
#define NOTE_FS2 93
#define NOTE_G2    98
#define NOTE_GS2 104
#define NOTE_A2    110
#define NOTE_AS2 117
#define NOTE_B2    123
#define NOTE_C3    131
#define NOTE_CS3 139
#define NOTE_D3    147
#define NOTE_DS3 156
#define NOTE_E3    165
```

```
#define NOTE_F3   175
#define NOTE_FS3 185
#define NOTE_G3   196
#define NOTE_GS3 208
#define NOTE_A3   220
#define NOTE_AS3 233
#define NOTE_B3   247
#define NOTE_C4   262
#define NOTE_CS4 277
#define NOTE_D4   294
#define NOTE_DS4 311
#define NOTE_E4   330
#define NOTE_F4   349
#define NOTE_FS4 370
#define NOTE_G4   392
#define NOTE_GS4 415
#define NOTE_A4   440
#define NOTE_AS4 466
#define NOTE_B4   494
#define NOTE_C5   523
#define NOTE_CS5 554
#define NOTE_D5   587
#define NOTE_DS5 622
#define NOTE_E5   659
#define NOTE_F5   698
#define NOTE_FS5 740
#define NOTE_G5   784
#define NOTE_GS5 831
#define NOTE_A5   880
#define NOTE_AS5 932
#define NOTE_B5   988
#define NOTE_C6   1047
#define NOTE_CS6 1109
#define NOTE_D6   1175
#define NOTE_DS6 1245
#define NOTE_E6   1319
#define NOTE_F6   1397
#define NOTE_FS6 1480
#define NOTE_G6   1568
```

```
#define NOTE_GS6 1661
#define NOTE_A6    1760
#define NOTE_AS6 1865
#define NOTE_B6    1976
#define NOTE_C7    2093
#define NOTE_CS7 2217
#define NOTE_D7    2349
#define NOTE_DS7 2489
#define NOTE_E7    2637
#define NOTE_F7    2794
#define NOTE_FS7 2960
#define NOTE_G7    3136
#define NOTE_GS7 3322
#define NOTE_A7`   3520
#define NOTE_AS7 3729
#define NOTE_B7    3951
#define NOTE_C8    4186
#define NOTE_CS8 4435
#define NOTE_D8    4699
#define NOTE_DS8 4978
```

shiftOut(dataPin, clockPin, bitOrder, value)

把資料傳給用來延伸數位輸出的暫存器,函式使用一個腳位表示資料、一個腳位表示時脈。bitOrder 用來表示位元間移動的方式(LSBFIRST 最低有效位元或是 MSBFIRST 最高有效位元),最後 value 會以 byte 形式輸出。此函式通常使用在延伸數位的輸出。

範例:

```
#define dataPin 8
#define clockPin 7
void setup()
{
shiftOut(dataPin, clockPin, LSBFIRST, 255);
}
void loop()
```

```
{    }
```

unsigned long pulseIn(Pin, value)

設定讀取腳位狀態的持續時間，例如使用紅外線、加速度感測器測得某一項數值時，在時間單位內不會改變狀態。

範例：

```
#define dataPin 8
#define pulsein 7
void setup()
{
Int time ;
time = pulsein(pulsein,HIGH); // 設定腳位 7 的狀態在時間單位內保持為 HIGH
}
void loop()
{    }
```

時間函式

➢ millis()
➢ micros()
➢ delay()
➢ delayMicroseconds()

控制與計算晶片執行期間的時間

unsigned long millis()

回傳晶片開始執行到目前的毫秒

範例:

```
int    lastTime ,duration;
void setup()
{
   lastTime = millis() ;
```

```
}
void loop()
{
  duration = -lastTime; //  表示自"lastTime"至當下的時間
}
```

delay(ms)

暫停晶片執行多少毫秒

範例:

```
void setup()
{
  Serial.begin(9600);
}
void loop()
{
  Serial.print(millis()) ;
  delay(500); //暫停半秒（500毫秒）
}
```

「毫」是 10 的負 3 次方的意思，所以「毫秒」就是 10 的負 3 次方秒，也就是 0.001 秒。

表 3 常用單位轉換表

符號	中文	英文	符號意義
p	微微	pico	10 的負 12 次方
n	奈	nano	10 的負 9 次方
u	微	micro	10 的負 6 次方
m	毫	milli	10 的負 3 次方
K	仟	kilo	10 的 3 次方
M	百萬	mega	10 的 6 次方
G	十億	giga	10 的 9 次方
T	兆	tera	10 的 12 次方

delay Microseconds(us)

暫停晶片執行多少微秒

範例:

```
void setup()
{
    Serial.begin(9600);
}
void loop()
{
    Serial.print(millis()) ;
    delayMicroseconds (1000); //暫停半秒（500 毫秒）
}
```

數學函式

- ➤ min()
- ➤ max()
- ➤ abs()
- ➤ constrain()
- ➤ map()
- ➤ pow()
- ➤ sqrt()

三角函式以及基本的數學運算

min(x, y)

回傳兩數之間較小者

範例：

```
#define sensorPin1 7
#define sensorPin2 8
void setup()
{
    int val;
```

```
    PinMode(sensorPin1,INPUT); // 將腳位 sensorPin1 (7) 定為輸入模式
    PinMode(sensorPin2,INPUT); // 將腳位 sensorPin2 (8) 定為輸入模式
    val = min(analogRead (sensorPin1), analogRead (sensorPin2)) ;
}
void loop()
{    }
```

max(x, y)

回傳兩數之間較大者

範例：

```
#define sensorPin1 7
#define sensorPin2 8
void setup()
{
  int val;
  PinMode(sensorPin1,INPUT); // 將腳位 sensorPin1 (7) 定為輸入模式
  PinMode(sensorPin2,INPUT); // 將腳位 sensorPin2 (8) 定為輸入模式
  val = max (analogRead (sensorPin1), analogRead (sensorPin2)) ;
}
void loop()
{    }
```

abs(x)

回傳該數的絕對值，可以將負數轉正數。

範例：

```
#define sensorPin1 7
void setup()
{
  int val;
  PinMode(sensorPin1,INPUT); // 將腳位 sensorPin (7) 定為輸入模式
    val = abs(analogRead (sensorPin1)-500);
        // 回傳讀值-500 的絕對值
```

```
}
void loop()
{      }
```

constrain(x, a, b)

判斷 x 變數位於 a 與 b 之間的狀態。x 若小於 a 回傳 a；介於 a 與 b 之間回傳 x 本身；大於 b 回傳 b

範例：

```
#define sensorPin1 7
#define sensorPin2 8
#define sensorPin 12
void setup()
{
  int val;
  PinMode(sensorPin1,INPUT); // 將腳位  sensorPin1 (7)  定為輸入模式
  PinMode(sensorPin2,INPUT); // 將腳位  sensorPin2 (8)  定為輸入模式
  PinMode(sensorPin,INPUT); // 將腳位  sensorPin (12)  定為輸入模式
  val = constrain(analogRead(sensorPin), analogRead (sensorPin1), analogRead
(sensorPin2)) ;
  // 忽略大於 255 的數
}
void loop()
{
}
```

map(value, fromLow, fromHigh, toLow, toHigh)

將 value 變數依照 fromLow 與 fromHigh 範圍，對等轉換至 toLow 與 toHigh 範圍。時常使用於讀取類比訊號，轉換至程式所需要的範圍值。

例如：

```
#define sensorPin1 7
#define sensorPin2 8
```

```
#define sensorPin 12
void setup()
{
  int val;
  PinMode(sensorPin1,INPUT); // 將腳位 sensorPin1 (7) 定為輸入模式
  PinMode(sensorPin2,INPUT); // 將腳位 sensorPin2 (8) 定為輸入模式
  PinMode(sensorPin,INPUT);  // 將腳位 sensorPin (12) 定為輸入模式
  val = map(analogRead(sensorPin), analogRead (sensorPin1), analogRead
(sensorPin2),0,100) ;
  // 將 analog0 所讀取到的訊號對等轉換至 100 – 200 之間的數值
}
void loop()
{      }
```

double pow(base, exponent)

回傳一個數(base)的指數(exponent)值。

範例：

```
int y=2;
double x = pow(y, 32); // 設定 x 為 y 的 32 次方
```

double sqrt(x)

回傳 double 型態的取平方根值。

範例：

```
int y=2123;
double x = sqrt (y);   // 回傳 2123 平方根的近似值
```

三角函式

➤　sin()
➤　cos()
➤　tan()

double sin(rad)

回傳角度（radians）的三角函式 sine 值。

範例：

```
int y=45;
double sine = sin (y);   // 近似值 0.70710678118654
```

double cos(rad)

回傳角度（radians）的三角函式 cosine 值。

範例：

```
int y=45;
double cosine = cos (y);   // 近似值 0.70710678118654
```

double tan(rad)

回傳角度（radians）的三角函式 tangent 值。

範例：

```
int y=45;
double tangent = tan (y);   // 近似值 1
```

亂數函式

- ➢ randomSeed()
- ➢ random()

本函數是用來產生亂數用途：

randomSeed(seed)

事實上在 Arduino 裡的亂數是可以被預知的。所以如果需要一個真正的亂數，可以呼叫此函式重新設定產生亂數種子。你可以使用亂數當作亂數的種子，以確保數字以隨機的方式出現，通常會使用類比輸入當作亂數種子，藉此可以產生與環境有關的亂數。

範例：

```
#define sensorPin 7
void setup()
{
randomSeed(analogRead(sensorPin)); // 使用類比輸入當作亂數種子
}
void loop()
{
}
```

long random(min, max)

回傳指定區間的亂數，型態為 long。如果沒有指定最小值，預設為 0。

範例：

```
#define sensorPin 7
long randNumber;
void setup(){
  Serial.begin(9600);
  // if analog input Pin sensorPin(7) is unconnected, random analog
  // noise will cause the call to randomSeed() to generate
  // different seed numbers each time the sketch runs.
  // randomSeed() will then shuffle the random function.
  randomSeed(analogRead(sensorPin));
}
void loop() {
  // print a random number from 0 to 299
  randNumber = random(300);
```

```
    Serial.println(randNumber);

    // print a random number from    0 to 100
    randNumber = random(0, 100);    // 回傳 0 – 99 之間的數字
    Serial.println(randNumber);
    delay(50);
}
```

通訊函式

你可以在許多例子中，看見一些使用序列埠與電腦交換資訊的範例，以下是函式解釋。

Serial.begin(speed)

你可以指定 Arduino 從電腦交換資訊的速率，通常我們使用 9600 bps。當然也可以使用其他的速度，但是通常不會超過 115,200 bps（每秒位元組）。

範例：

```
void setup() {
   Serial.begin(9600);          // open the serial port at 9600 bps:
}
void loop() {
  }
```

Serial.print(data)
Serial.print(data, 格式字串(encoding))

經序列埠傳送資料，提供編碼方式的選項。如果沒有指定，預設以一般文字傳送。

範例：

```
int x = 0;        // variable

void setup() {
  Serial.begin(9600);          // open the serial port at 9600 bps:
}

void loop() {
  // print labels
  Serial.print("NO FORMAT");          // prints a label
  Serial.print("\t");                 // prints a tab
  Serial.print("DEC");
  Serial.print("\t");
  Serial.print("HEX");
  Serial.print("\t");
  Serial.print("OCT");
  Serial.print("\t");
  Serial.print("BIN");
  Serial.print("\t");
}
```

Serial.println(data)

Serial.println(data, ,格式字串(encoding))

　　與 Serial.print()相同，但會在資料尾端加上換行字元（　）。意思如同你在鍵盤上打了一些資料後按下 Enter。

　　範例：

```
int x = 0;        // variable
void setup() {
  Serial.begin(9600);          // open the serial port at 9600 bps:
}
void loop() {
  // print labels
  Serial.print("NO FORMAT");          // prints a label
  Serial.print("\t");                 // prints a tab
  Serial.print("DEC");
```

```
Serial.print("\t");
Serial.print("HEX");
Serial.print("\t");
Serial.print("OCT");
Serial.print("\t");
Serial.print("BIN");
Serial.print("\t");

for(x=0; x< 64; x++){        // only part of the ASCII chart, change to suit
    // print it out in many formats:
    Serial.print(x);             // print as an ASCII-encoded decimal - same as "DEC"
    Serial.print("\t");       // prints a tab
    Serial.print(x, DEC);    // print as an ASCII-encoded decimal
    Serial.print("\t");       // prints a tab
    Serial.print(x, HEX);    // print as an ASCII-encoded hexadecimal
    Serial.print("\t");       // prints a tab
    Serial.print(x, OCT);    // print as an ASCII-encoded octal
    Serial.print("\t");       // prints a tab
    Serial.println(x, BIN);   // print as an ASCII-encoded binary
    //                  then adds the carriage return with "println"
    delay(200);                  // delay 200 milliseconds
}
Serial.println("");            // prints another carriage return

}
```

格式字串(encoding)

Arduino 的 print()和 println()，在列印內容時，可以指定列印內容使用哪一種格式列印，若不指定，則以原有內容列印。

列印格式如下：

1. BIN(二進位，或以 2 為基數)，

2. OCT(八進制，或以 8 為基數)，

3. DEC(十進位，或以 10 為基數)，

4.   HEX(十六進位，或以 16 為基數)。

使用範例如下：

● Serial.print(78,BIN)輸出為"1001110"

● Serial.print(78,OCT)輸出為"116"

● Serial.print(78,DEC)輸出為"78"

● Serial.print(78,HEX)輸出為"4E"

對於浮點型數位，可以指定輸出的小數數位。例如

● Serial.println(1.23456,0)輸出為"1"

● Serial.println(1.23456,2)輸出為"1.23"

● Serial.println(1.23456,4)輸出為"1.2346"

Print & Println 列印格式(printformat01)

```
/*
使用 for 迴圈列印一個數字的各種格式。
*/
int x = 0;       // 定義一個變數並賦值

void setup() {
   Serial.begin(9600);          // 打開串口傳輸，並設置串列傳輸速率為 9600
}

void loop() {
   ///列印標籤
   Serial.print("NO FORMAT");          // 列印一個標籤
   Serial.print("\t");                 // 列印一個轉義字元

   Serial.print("DEC");
```

```
    Serial.print("\t");

    Serial.print("HEX");
    Serial.print("\t");

    Serial.print("OCT");
    Serial.print("\t");

    Serial.print("BIN");
    Serial.print("\t");

    for(x=0; x< 64; x++){        // 列印 ASCII 碼表的一部分，修改它的格式得到需
要的內容

        //  列印多種格式：
        Serial.print(x);         // 以十進位格式將 x 列印輸出 - 與 "DEC"相同
        Serial.print("\t");      // 橫向跳格

        Serial.print(x, DEC);    // 以十進位格式將 x 列印輸出
        Serial.print("\t");      // 橫向跳格

        Serial.print(x, HEX);    // 以十六進位格式列印輸出
        Serial.print("\t");      // 橫向跳格

        Serial.print(x, OCT);    // 以八進制格式列印輸出
        Serial.print("\t");      // 橫向跳格

        Serial.println(x, BIN);  // 以二進位格式列印輸出
        //                                    然後用 "println"列印一個回車
        delay(200);              // 延時 200ms
    }
    Serial.println("");          // 列印一個空字元，並自動換行
}
```

int Serial.available()

回傳有多少位元組（bytes）的資料尚未被 read()函式讀取，如果回傳值是 0 代表所有序列埠上資料都已經被 read()函式讀取。

範例：

```
int incomingByte = 0;    // for incoming serial data
  void setup() {
          Serial.begin(9600);          // opens serial port, sets data rate to 9600 bps
}
  void loop() {
          // send data only when you receive data:
          if (Serial.available() > 0) {
                  // read the incoming byte:
                  incomingByte = Serial.read();
                  // say what you got:
                  Serial.print("I received: ");
                  Serial.println(incomingByte, DEC);
          }
}
```

int Serial.read()

以 byte 方式讀取 1byte 的序列資料

範例：

```
int incomingByte = 0;    // for incoming serial data
void setup() {
   Serial.begin(9600);          // opens serial port, sets data rate to 9600 bps
}
void loop() {
   // send data only when you receive data:
   if (Serial.available() > 0) {
      // read the incoming byte:
      incomingByte = Serial.read();
      // say what you got:
      Serial.print("I received: ");
      Serial.println(incomingByte, DEC);
```

```
    }
}
```

int Serial.write()

以 byte 方式寫入資料到序列

範例：

```
void setup(){
   Serial.begin(9600);
}
void loop(){
   Serial.write(45); // send a byte with the value 45
     int bytesSent = Serial.write("hello Arduino , I am a beginner in the Arduino
world");
}
```

Serial.flush()

有時候因為資料速度太快，超過程式處理資料的速度，你可以使用此函式清除
緩衝區內的資料。經過此函式可以確保緩衝區(buffer)內的資料都是最新的。

範例：

```
void setup(){
   Serial.begin(9600);
}
void loop(){
   Serial.write(45); // send a byte with the value 45
     int bytesSent = Serial.write("hello Arduino , I am a beginner in the Arduino
world");
       Serial.flush();
   }
```

系統函式

Arduino 開發版也提供許多硬體相關的函式：

系統 idle 函式

使用硬體 idle 功能，可以讓 Arduin 進入睡眠狀態，連單晶片都可以進入睡眠狀態，但使用本功能需要使用外掛函式 Enerlib 函式庫，讀者可以到 Arduino 官網：http://playground.arduino.cc/Code/Enerlib，下載其函式庫安裝，或到本書範例檔：https://github.com/brucetsao/arduino_RFProgramming，下載相關函式與範例。

ATMega328 微控器具有六種睡眠模式，底下是依照「省電情況」排列的睡眠模式名稱，以及 Enerlib（註：Energy 和 Library，即：「能源」和「程式庫」的縮寫）程式庫的五道函數指令對照表，排越後面越省電。「消耗電流」欄位指的是 ATmega328 處理器本身，而非整個控制板(趙英傑, 2013, 2014)。

表 4 ATMega328 微控器六種睡眠模式

睡眠模式	Energy 指令	中文直譯	消耗電流
Idle	Idle()	閒置	15mA
ADC Noise Reduc-tion	SleepADC()	類比數位轉換器降低雜訊	6.5mA
Power-save	PowerSave()	省電	1.62mA
Standby	Standby()	待機	1.62mA
Extended Standby		延長待機	0.84mA
Power-down	PowerDown()	斷電	0.36mA

微控器內部除了中央處理器（CPU）， 還有記憶體、類比數位轉換器、序列通訊…等模組。越省電的模式，仍在運作中的模組就越少。

例如，在＂Power-Down＂（電源關閉）睡眠模式之下，微控器僅剩下**外部中斷**和**看門狗計時器**（**Watchdog Timer**）仍持續運作。而在 Idle 睡眠模式底下，SPI, UART（也就是序列埠）、計時器、類比數位轉換器等，仍持續運作，只有中央處理器和快閃記憶體（Flash）時脈訊號被停止。

時脈訊號就像心跳一樣，一旦停止時脈訊號，相關的元件也隨之暫停。各種睡眠模式的詳細說明，請參閱下表。

表 5 Active Clock Domains and Wake-up Sources in the Different Sleep Modes

Active Clock Domains and Wake-up Sources in the Different Sleep Modes.

Sleep Mode	Active Clock Domains					Oscillators		Wake-up Sources							
	clk$_{CPU}$	clk$_{FLASH}$	clk$_{IO}$	clk$_{ADC}$	clk$_{ASY}$	Main Clock Source Enabled	Timer Oscillator Enabled	INT1, INT0 and Pin Change	TWI Address Match	Timer2	SPM/EEPROM Ready	ADC	WDT	Other I/O	Software BOD Disable
Idle			X	X	X	X	X$^{(2)}$	X	X	X	X	X	X	X	
ADC Noise Reduction			X	X	X	X	X$^{(2)}$	X$^{(3)}$	X	X$^{(2)}$	X	X	X		
Power-down								X$^{(3)}$	X				X		X
Power-save					X		X$^{(2)}$	X$^{(3)}$	X	X			X		X
Standby$^{(1)}$						X		X$^{(3)}$	X				X		X
Extended Standby					X$^{(2)}$	X	X$^{(2)}$	X$^{(3)}$	X	X			X		X

Notes: 1. Only recommended with external crystal or resonator selected as clock source.
2. If Timer/Counter2 is running in asynchronous mode.
3. For INT1 and INT0, only level interrupt.

資料來源：ATmega328 微控器的資料手冊，第 39 頁，「Power Management and Sleep Modes（電源管理與睡眠模式）」單元

(http://www.atmel.com/images/doc8161.pdf)

範例：

Arduin 進入睡眠狀態範例(
/*
Enerlib: easy-to-use wrapper for AVR's Sleep library.

```
Example showing how to enter in Idle mode and exit from it with INT0.
*/


#include <Enerlib.h>

Energy energy;

void INT0_ISR(void)
{

  /*
  The WasSleeping function will return true if Arduino
  was sleeping before the IRQ. Subsequent calls to
  WasSleeping will return false until Arduino reenters
  in a low power state. The WasSleeping function should
  only be called in the ISR.
  */
  if (energy.WasSleeping())
  {
    /*
    Arduino was waked up by IRQ.

    If you shut down external peripherals before sleeping, you
    can reinitialize them here. Look on ATMega's datasheet for
    hardware limitations in the ISR when microcontroller just
    leave any low power state.
    */
  }
  else
  {
    /*
    The IRQ happened in awake state.

    This code is for the "normal" ISR.
    */
  }
}
```

```
void setup()
{
    Serial.begin(9600);
    Serial.println("Program Start") ;
  attachInterrupt(0, INT0_ISR, LOW);
  /*
  Pin 2 will be the "wake button". Due to uC limitations,
  it needs to be a level interrupt.
  For experienced programmers:
    ATMega's datasheet contains information about the rest of
    wake up sources. The Extended Standby is not implemented.
  */

  Serial.println("Now I am Sleeping") ;
  delay(500);
  energy.Idle();
}

void loop()
{

    Serial.println("I am waken ") ;
  delay(1000);
}
```

attachInterrupt(插斷)

　　當開發者攥寫程式時，在 loop()程式段之中，攥寫許多大量的程式碼，並且重覆的執行，當我們需要在某些時後去檢查某一樣硬體，如按鈕、讀卡機、RFID、鍵盤、滑鼠等周邊裝置，若這些檢查、讀取該周邊的函式，寫在 loop()程式段之中，則必需每一個迴圈都必需耗時去檢查，不但造成程式不順暢，還會錯失讀取這些檢查、讀取該周邊的函式的時機。

這時後我們就需要用到這些 Arduino 開發板外部插斷接腳，由於 Arduino 開發板使用外部插斷接腳，不同開發板其接腳都不太相同，我們可以參考表 6 之 Arduino 開發板外部插斷接腳對照表。

表 6 Arduino 開發板外部插斷接腳對照表

Board	int.0	int.1	int.2	int.3	int.4	int.5
Uno, Ethernet	2	3				
Mega2560	2	3	21	20	19	18
Leonardo	3	2	0	1	7	

attachInterrupt(第幾號外部插斷, 執行之函式名稱, LOW/HIGH);

參數解說：

- 第一個參數：使用那一個外部插斷，其接腳請參考參考表 6 之 Arduino 開發板外部插斷接腳對照表。

- 第二個參數：為執行之函式名稱；在 Arduino 程式區自行定義一個若使用插斷後，執行的函式名稱

- 第三個參數：驅動外部硬體插斷所使用的電位， HIGH 表高電位，LOW 表低電位

範例：

```
外部插斷測式程式(IRQTest)
void setup() {
  // put your setup code here, to run once:

  Serial.begin(9600);
   Serial.println("Program Start") ;
  attachInterrupt(0, TheButtonPressed, LOW);
}
```

```
void loop() {
  // put your main code here, to run repeatedly:
    Serial.print("now program run in loop()");
}

void TheButtonPressed()
{
  Serial.println("The Button is pressed by user") ;
}
```

章節小結

　　本章節概略的介紹 Arduino 程式攥寫的語法、函式等介紹，接下來就是介紹本書主要的內容，讓我們視目以待。

CHAPTER

LCD 顯示螢幕

由於許多電子線路必須將內部的狀態資訊顯示到可見的裝置,供使用者讀取資訊,方能夠繼續使用,所以我們必須提供一個可以顯示電子線路內在資訊的顯示介面,通常我們使用一個獨立的顯示螢幕,使我們的設計更加完整。

LCD 1602

為了達到這個目的,先行介紹 Arduino 開發板常用 LCD 1602 ,常見的 LCD 1602 是和日立的 HD44780[7] 相容的 2x16 LCD ,可以顯示兩行資訊,每行 16 個字元,它可以顯示英文字母、希臘字母、標點符號以及數學符號。

除了顯示資訊外,它還有其它功能,包括資訊捲動(往左和往右捲動)、顯示游標和 LED 背光的功能,但是有一些廠商為了降低售價,取消其 LED 背光的功能。

如下圖所示,大部分的 LCD 1602 都配備有背光裝置,所以大部份具有 16 個腳位,可以參考下表所示,可以更深入了解其接腳功能與定義(曹永忠, 許智誠, & 蔡英德, 2015a, 2015b, 2015f, 2015g, 2015h, 2015i, 2015j, 2015l; 曹永忠, 許碩芳, 許智誠, & 蔡英德, 2015a):

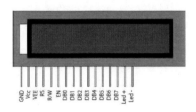

---

[7] Hitachi HD44780 LCD controller is one of the most common dot matrix liquid crystal display (LCD) display controllers available. Hitachi developed the microcontroller specifically to drive alphanumeric LCD display with a simple interface that could be connected to a general purpose microcontroller or microprocessor

圖 31 LCD1602 接腳

表 7 LCD1602 接腳說明表

接腳	接腳說明	接腳名稱
1	Ground (0V)	接地 (0V)
2	Supply voltage; 5V (4.7V – 5.3V)	電源 (+5V)
3	Contrast adjustment; through a variable resistor	螢幕對比(0-5V), 可接一顆 1k 電阻到地線，或使用可變電阻調整適當的對比(請參考分壓線路) **請參考下圖分壓線路**
4	Selects command register when low; and data register when high	Register Select: 1: D0 – D7 當作資料解釋 0: D0 – D7 當作指令解釋
5	Low to write to the register; High to read from the register	Read/Write mode: 1: 從 LCD 讀取資料 0: 寫資料到 LCD 因為很少從 LCD 這端讀取資料，可將此腳位接地以節省 I/O 腳位。 *****若不使用此腳位，請接地**
6	Sends data to data pins when a high to low pulse is given	Enable
7	8-bit data pins	Bit 0 LSB
8		Bit 1
9		Bit 2
10		Bit 3
11		Bit 4
12		Bit 5
13		Bit 6
14		Bit 7 MSB
15	Backlight V$_{cc}$ (5V)	背光(串接 330 R 電阻到電源)
16	Backlight Ground (0V)	背光(GND)

資料來源：(Guangzhou_Tinsharp_Industrial_Corp._Ltd., 2013)

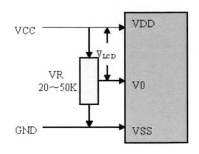

圖 32 LCD1602 對比線路(分壓線路)

　　若讀者要調整 LCD 1602 顯示文字的對比，請參考分壓線路，不可以直接連接
+5V 或接地，避免 LCD 1602 或 Arduino 開發板燒毀。

　　為了讓實驗更順暢進行，先行介紹 LCD1602
(Guangzhou_Tinsharp_Industrial_Corp._Ltd., 2013) ，我們參考下圖所示，如何將 LCD
1602 與 Arduino 開發板連接起來，並可以參考下圖之接線圖，將 LCD 1602 與 Arduino
開發板進行實體線路連接，參考附錄中，LCD 1602 函式庫 單元，可以見到 LCD
1602 常用的函式庫(LiquidCrystal Library,參考網址：
http://arduino.cc/en/Reference/LiquidCrystal)，若讀者希望對 LCD 1602 有更深入的了
解，可以參考『Ameba 空氣粒子感測裝置設計與開發(MQTT 篇)):Using Ameba to
Develop a PM 2.5 Monitoring Device to MQTT』(曹永忠, 許智誠, & 蔡英德, 2015d)、
『Ameba 空气粒子感测装置设计与开发(MQTT 篇):Using Ameba to Develop a PM 2.5
Monitoring Device to MQTT』(曹永忠, 許智誠, & 蔡英德, 2015c)等書附錄中 LCD
1602 原廠資料(Guangzhou_Tinsharp_Industrial_Corp._Ltd., 2013)，相信會有更詳細的資
料介紹。

　　LCD 1602 有 4-bit 和 8-bit 兩種使用模式，使用 4-bit 模式主要的好處是節省
I/O 腳位，通訊的時候只會用到 4 個高位元 (D4-D7)，D0-D3 這四支腳位可以不
用接。每個送到 LCD 1602 的資料會被分成兩次傳送 – 先送 4 個高位元資料，
然後才送 4 個低位元資料。

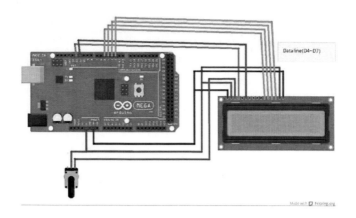

圖 33 LCD 1602 接線示意圖

使用工具 by Fritzing (Fritzing.org., 2013)

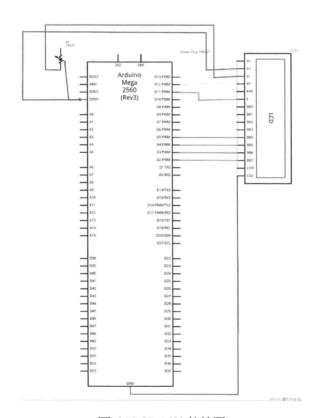

圖 34 LCD 1602 接線圖

使用工具 by Fritzing (Fritzing.org., 2013)

我們參考 Arduino 官方網站 http://arduino.cc/en/Reference/LiquidCrystal ，其連接 LCD 1602 範例程式，可以了解 Arduino 如何驅動 LCD 1602 顯示器：

表 8 LCD 1602 接腳範例圖

接腳	接腳說明	接腳名稱
1	Ground (0V)	接地 (0V)
2	Supply voltage; 5V (4.7V – 5.3V)	電源 (+5V)
3	Contrast adjustment; through a variable resistor	螢幕對比(0-5V), 可接一顆 1k 電阻，或使用可變電阻調整適當的對比**(請參考分壓線路)(曹永忠, 許智誠, et al., 2015c, 2015d)**
4	Selects command register when low; and data register when high	Arduino digital output pin 5
5	Low to write to the register; High to read from the register	Arduino digital output pin 6
6	Sends data to data pins when a high to low pulse is given	Arduino digital output pin 7
7	Data D0	Arduino digital output pin 30
8	Data D1	Arduino digital output pin 32
9	Data D2	Arduino digital output pin 34
10	Data D3	Arduino digital output pin 36
11	Data D4	Arduino digital output pin 38
12	Data D5	Arduino digital output pin 40
13	Data D6	Arduino digital output pin 42
14	Data D7	Arduino digital output pin 44
15	Backlight V_{cc} (5V)	背光(串接 330 R 電阻到電源)
16	Backlight Ground (0V)	背光(GND)

如下表所示，為 LiquidCrystal LCD 1602 測試程式，請讀者鍵入Ｓｋｅｔｃｈ ＩＤＥ軟體，編譯完成後上傳到開發版進行測試。

表 9 LiquidCrystal LCD 1602 測試程式

LiquidCrystal LCD 1602 測試程式(lcd1602_hello)

```
/*
LiquidCrystal Library - Hello World

Use a 16x2 LCD display The LiquidCrystal
library works with all LCD displays that are compatible with the
Hitachi HD44780 driver.
This sketch prints "Hello World!" to the LCD
and shows the time.
*/
// include the library code:
#include <LiquidCrystal.h>
// initialize the library with the numbers of the interface pins
LiquidCrystal lcd(5,6,7,38,40,42,44);    //ok
void setup() {
// set up the LCD's number of columns and rows:
lcd.begin(16, 2);
// Print a message to the LCD.
lcd.print("hello, world!");
}
void loop() {
lcd.setCursor(0, 1);
lcd.print(millis()/1000);    }
```

如下表所示，為 LiquidCrystal LCD 1602 測試程式二，請讀者鍵入Ｓｋｅｔｃｈ
ＩＤＥ軟體，編譯完成後上傳到開發版進行測試。

表 10 LiquidCrystal LCD 1602 測試程式二

LiquidCrystal LCD 1602 測試程式(lcd1602_mills)
#include <LiquidCrystal.h>   /* LiquidCrystal display with: LiquidCrystal(rs, enable, d4, d5, d6, d7)   LiquidCrystal(rs, rw, enable, d4, d5, d6, d7)   LiquidCrystal(rs, enable, d0, d1, d2, d3, d4, d5, d6, d7)

```
    LiquidCrystal(rs, rw, enable, d0, d1, d2, d3, d4, d5, d6, d7)
    R/W Pin Read = LOW / Write = HIGH     // if No pin connect RW , please leave
R/W Pin for Low State


    Parameters
    */
    LiquidCrystal lcd(5,6,7,38,40,42,44);      //ok

    void setup()
    {
      Serial.begin(9600);
      Serial.println("start LCM 1604");
      //   pinMode(11,OUTPUT);
      //   digitalWrite(11,LOW);
      lcd.begin(16, 2);
      // 設定 LCD 的行列數目 (16 x 2)  16  行2  列
      lcd.setCursor(0,0);
      // 列印 "Hello World" 訊息到 LCD 上
      lcd.print("hello, world!");
      Serial.println("hello, world!");
    }


    void loop()
    {
      // 將游標設到  第一行,   第二列
      // (注意:   第二列第五行,因為是從 0 開始數起):
      lcd.setCursor(5, 2);
      // 列印 Arduino 重開之後經過的秒數
      lcd.print(millis()/1000);
      Serial.println(millis()/1000);
      delay(200);
    }
```

LCD 2004

為了達到這個目的，先行介紹 Arduino 開發板常用 LCD 2004 ，常見的 LCD 2004 是 SPLC780D 驅動 IC，類似 LCD 1602 驅動 IC HD44780(可參考 LCD 1602 一章) ，可以顯示四行資訊，每行 20 個字元，它可以顯示英文字母、希臘字母、標點符號以及數學符號。

除了顯示資訊外，它還有其它功能，包括資訊捲動(往左和往右捲動)、顯示游標和 LED 背光的功能，但是有一些廠商為了降低售價，取消其 LED 背光的功能。

如下圖所示，大部分的 LCD 2004 都配備有背光裝置，所以大部份具有 16 個腳位，因為其需要占住 RS/Enable 兩個控制腳位與 D0~D7 八個資料腳位或 D4~D7 四個資料腳位，所以有許多廠商開發出 I2C 的版本，可參考下下圖所示，可以省下至少四個以上的腳位。對於接腳資料，可以參考下表所示，LCD 2004 的腳位與 LCD 1602 腳位相容，讀者可以更深入了解其接腳功能與定義：

圖 35 LCD 2004 外觀圖

圖 36 LCD 2004 I2C 版本

表 11 LCD 2004 接腳說明表

接腳	接腳說明	接腳名稱
1	Ground (0V)	接地 (0V)
2	Supply voltage; 5V (4.7V － 5.3V)	電源 (+5V)
3	Contrast adjustment; through a variable resistor	螢幕對比(0-5V), 可接一顆 1k 電阻，或使用可變電阻調整適當的對比(請參考分壓線路) **請參考下圖分壓線路(曹永忠, 許智誠, et al., 2015c, 2015d)**
4	Selects command register when low; and data register when high	Register Select: 　1: D0 － D7 當作資料解釋 　0: D0 － D7 當作指令解釋
5	Low to write to the register; High to read from the register	Read/Write mode: 　1: 從 LCD 讀取資料 　0: 寫資料到 LCD 因為很少從 LCD 這端讀取資料，可將此腳位接地以節省 I/O 腳位。 ***若不使用此腳位，請接地**
6	Sends data to data pins when a high to low pulse is given	Enable
7		Bit 0 LSB
8		Bit 1
9		Bit 2
10		Bit 3
11	8-bit data pins	Bit 4
12		Bit 5
13		Bit 6
14		Bit 7 MSB
15	Backlight Vcc (5V)	背光(串接 330 R 電阻到電源)
16	Backlight Ground (0V)	背光(GND)

資料來源：SHENZHEN EONE ELECTRONICS CO.,LTD，下載位址：

ftp://imall.iteadstudio.com/IM120424018_EONE_2004_Characters_LCD/SPE_IM120424018_EONE_2004_Ch

aracters_LCD.pdf

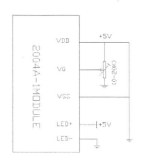

圖 37 LCD2004 對比線路(分壓線路)

　　若讀者要調 LCD 2004 顯示文字的對比，請參考分壓線路，不可以直接連接+5V
或接地，避免 LCD 2004 或 Arduino 開發板損壞。

　　為了讓實驗更順暢進行，先行介紹 LCD 2004，我們參考下表所示，如何將 LCD
2004 與 Arduino 開發板連接起來(與 LCD 1602 相同接法)，將 LCD 2004 與 Arduino
開發板進行實體線路連接，參考本文 LCD 1602 函式庫單元(LCD 2004 函式庫與函
式庫共用)，可以見到 LCD 2004(相容於 LCD 1602)常用的函式庫(LiquidCrystal Library,
參考網址：http://arduino.cc/en/Reference/LiquidCrystal)，若讀者希望對 LCD 2004 有更
深入的了解，可以參考附錄中 LCD 2004 原廠資料(SHENZHEN EONE ELECTRONICS
CO.,LTD，下載位址：
ftp://imall.iteadstudio.com/IM120424018_EONE_2004_Characters_LCD/SPE_IM120424018
_EONE_2004_Characters_LCD.pdf)，相信會有更詳細的資料介紹。

　　LCD 2004 有 4-bit 和 8-bit 兩種使用模式，使用 4-bit 模式主要的好處是節省
I/O 腳位，通訊的時候只會用到 4 個高位元 (D4-D7)，D0-D3 這四支腳位可以不
用接。每個送到 LCD 2004 的資料會被分成兩次傳送 – 先送 4 個高位元資料，然
後才送 4 個低位元資料。

　　我們參考 Arduino 官方網站 http://arduino.cc/en/Reference/LiquidCrystal ，其連
接 LCD 1602 範例程式，可以見到 Arduino 如何驅動 LCD 2004 顯示器：

表 12 LCD LCD 2004 範例桩腳圖

接腳	接腳說明	接腳名稱
1	Ground (0V)	接地 (0V)
2	Supply voltage; 5V (4.7V – 5.3V)	電源 (+5V)
3	Contrast adjustment; through a variable resistor	螢幕對比(0-5V), 可接一顆 1k 電阻, 或使用可變電阻調整適當的對比 (請參考分壓線路)(曹永忠, 許智誠, et al., 2015c, 2015d)
4	Selects command register when low; and data register when high	Arduino digital output pin 5
5	Low to write to the register; High to read from the register	Arduino digital output pin 6
6	Sends data to data pins when a high to low pulse is given	Arduino digital output pin 7
7	Data D0	Arduino digital output pin 30
8	Data D1	Arduino digital output pin 32
9	Data D2	Arduino digital output pin 34
10	Data D3	Arduino digital output pin 36
11	Data D4	Arduino digital output pin 38
12	Data D5	Arduino digital output pin 40
13	Data D6	Arduino digital output pin 42
14	Data D7	Arduino digital output pin 44
15	Backlight V_{CC} (5V)	背光(串接 330 R 電阻到電源)
16	Backlight Ground (0V)	背光(GND)

如下表所示，為 LiquidCrystal LCD 2004 測試程式，請讀者鍵入Ｓｋｅｔｃｈ ＩＤＥ軟體，編譯完成後上傳到開發版進行測試。

表 13 LiquidCrystal LCD 2004 測試程式

LiquidCrystal LCD 2004 測試程式(lcd2004_hello)
#include <LiquidCrystal.h>
/* LiquidCrystal display with:

```
LiquidCrystal(rs, enable, d4, d5, d6, d7)
LiquidCrystal(rs, rw, enable, d4, d5, d6, d7)
LiquidCrystal(rs, enable, d0, d1, d2, d3, d4, d5, d6, d7)
LiquidCrystal(rs, rw, enable, d0, d1, d2, d3, d4, d5, d6, d7)
R/W Pin Read = LOW / Write = HIGH      // if No pin connect RW , please leave R/W
Pin for Low State

Parameters
*/

LiquidCrystal lcd(5,6,7,38,40,42,44);      //ok

void setup()
{
   Serial.begin(9600);
   Serial.println("start LCM2004");
//   pinMode(11,OUTPUT);
//   digitalWrite(11,LOW);
lcd.begin(20, 4);
// 設定 LCD 的行列數目 (4 x 20)
 lcd.setCursor(0,0);
   // 列印 "Hello World" 訊息到 LCD 上
lcd.print("hello, world!");
   Serial.println("hello, world!");
}

void loop()
{
// 將游標設到 column 0, line 1
// (注意: line 1 是第二行(row)，因為是從 0 開始數起):
lcd.setCursor(0, 1);
// 列印 Arduino 重開之後經過的秒數
lcd.print(millis()/1000);
  Serial.println(millis()/1000);
delay(200);
}
```

如下表所示，為 LiquidCrystal LCD 2004 測試程式二，請讀者鍵入Ｓｋｅｔｃｈ
ＩＤＥ軟體，編譯完成後上傳到開發版進行測試。

表 14　LiquidCrystal LCD 2004 測試程式二

LiquidCrystal LCD 2004 測試程式(lcd2004_mills)

```
#include <LiquidCrystal.h>

/* LiquidCrystal display with:

 LiquidCrystal(rs, enable, d4, d5, d6, d7)
 LiquidCrystal(rs, rw, enable, d4, d5, d6, d7)
 LiquidCrystal(rs, enable, d0, d1, d2, d3, d4, d5, d6, d7)
 LiquidCrystal(rs, rw, enable, d0, d1, d2, d3, d4, d5, d6, d7)
 R/W Pin Read = LOW / Write = HIGH     // if No pin connect RW , please leave R/W
Pin for Low State

 Parameters
 */
LiquidCrystal lcd(5,6,7,38,40,42,44);     //ok
//
void setup()
{
  Serial.begin(9600);
  Serial.println("start LCM2004");
  //   pinMode(11,OUTPUT);
  //   digitalWrite(11,LOW);
  lcd.begin(20, 4);
  // 設定 LCD 的行列數目 (16 x 2)   16   行 2   列
  lcd.setCursor(0,0);
  // 列印 "Hello World" 訊息到 LCD 上
  lcd.print("hello, world!");
  Serial.println("hello, world!");
}

void loop()
{
```

```
// 將游標設到　第一行,　第二列
// (注意:　第二列第五行,因為是從 0 開始數起):
lcd.setCursor(5, 2);
// 列印 Arduino 重開之後經過的秒數
lcd.print(millis()/1000);
    Serial.println(millis()/1000);
    delay(200);
  }
```

LCD 1602 I²C 版

由上節看到,LCD1602 顯示模組共有 16 個腳位,去掉背光電源,電力,對白訊號等五條線,還有 11 個腳位需要接,對於微小的開發板,如 Pro Mini、Arduino Atiny、Arduino LilyPads...等,這樣的腳位數,似乎太多了,所以筆者介紹 LCD 1602 I²C 版(如下圖所示)。

(a). LCD1602 正面圖　　　(b). LCD 1602 I²C 轉接板

(c).LCD 1602 I²C 板

圖 38 LCD 1602 I²C 板

如下圖所示,其實 LCD 1602 I2C 板是由標準 LCD 1602(如上圖.(a)所示),加上 LCD 1602 I2C 轉接板(如上圖.(b)所示),所組合出來的 LCD 1602 I2C 板(如上圖.(c)所示),讀者可以先買標準 LCD 1602(如上圖.(a)所示),有需要的時後在買轉接板(如上圖.(b)所示),就可以組合成如下圖所示的成品。

圖 39 LCD 1602 I2C 零件表

圖 40 LCD 1602 I2C 組合圖

為了讓實驗更順暢進行，先參考下圖所示之 LCD 1602 I2C 接腳表，將 LCD
1602 I2C 板與 Arduino 開發板進行實體線路連接，參考本文 LCD 1602 函式庫 單
元，可以見到 LCD 1602 I2C 常用的函式庫(LiquidCrystal Library,參考網址：
http://arduino.cc/en/Reference/LiquidCrystal   ，
http://playground.arduino.cc/Code/LCDi2c   )。

表 15 LiquidCrystal Library API 相容表

Library	Displays Supported	Verified API	Connection
LCDi2cR	Robot-Electronics	Y	i2c
LCDi2cW	web4robot.com	Y	i2c
LiquidCrystal	Generic HitachiHD44780	P	4, 8 bit

LiquidCrystal_I2C	PCF8574drivingHD44780	Y	I2C
LCDi2cNHD	NewHavenDisplayI2CMode	Y	i2c
ST7036 Lib	GenericST7036LCD controller	Y	i2c

資料來源：Arduino 官網：http://playground.arduino.cc/Code/LCDAPI

由於不同種類的 Arduino 開發板，其 I2C/ TWI 接腳也略有不同，所以讀者可以參考下表所示之 Arduino 開發板 I2C/ TWI 接腳表，在根據下下表所示之 LCD 1602 I2C 測試程式的內容，進行硬體接腳的修正，至於軟體部份，Arduino 軟體原始碼的部份，則不需要修正。

表 16 Arduino 開發板 I2C/ TWI 接腳表

開發板種類	I2C/ TWI 接腳表
Uno, Ethernet	A4 (SDA), A5 (SCL)
Mega2560	20 (SDA), 21 (SCL)
Leonardo	2 (SDA), 3 (SCL)
Due	20 (SDA), 21 (SCL),SDA1,SCL1

我們參考 Arduino 官方網站 http://arduino.cc/en/Reference/LiquidCrystal ，其連接 LCD 1602 範例程式，可以了解 Arduino 如何驅動 LCD 1602 顯示器：

表 17 LCD 1602 I2C 接腳表

接腳	接腳說明	接腳名稱
1	Ground (0V)	接地 (0V) Arduino GND
2	Supply voltage; 5V (4.7V － 5.3V)	電源 (+5V) Arduino +5V
3	SDA	Arduino digital Pin20(SDA)
4	SCL	Arduino digital Pin21(SCL)

如下表所示，為 LCD 1602 I2C 測試程式，請讀者鍵入Ｓｋｅｔｃｈ　ＩＤＥ

軟體，編譯完成後上傳到開發版進行測試。

表 18 LCD 1602 I2C 測試程式

LCD 1602 I2C 測試程式(lcd1602_I2C_mill)

```
//Compatible with the Arduino IDE 1.0
//Library version:1.1
#include <Wire.h>
#include <LiquidCrystal_I2C.h>

LiquidCrystal_I2C lcd(0x27, 16, 2); // set the LCD address to 0x27 for a 16 chars and 2
line display

void setup()
{
  lcd.init();                            // initialize the lcd

  // Print a message to the LCD.
  lcd.backlight();
  lcd.print("Hello, world!");
}

void loop()
{
  // 將游標設到   第一行，  第二列
  // (注意:   第二列第五行，因為是從 0 開始數起):
  lcd.setCursor(5, 1);
  // 列印 Arduino 重開之後經過的秒數
  lcd.print(millis() / 1000);
  Serial.println(millis() / 1000);
  delay(200);
}
```

讀者也可以在筆者 YouTube 頻道

(https://www.youtube.com/user/UltimaBruce)中，在網址

https://www.youtube.com/watch?v=GXAplXXnVn8&feature=youtu.be，看到本次實驗-

LCD 1602 I2C 測試程式結果畫面。

　　當然、如下圖所示，我們可以看到 Arduino 在 LCD 1602 畫面上顯示文字情形。

圖 41 LCD 1602 I2C 測試程式結果畫面

LCD 2004 I²C 版

　　由上節看到，LCD2004 顯示模組共有 16 個腳位，去掉背光電源，電力，對白
訊號等五條線，還有 11 個腳位需要接，對於微小的開發板，如 Pro Mini、Arduino
Atiny、Arduino LilyPads...等，這樣的腳位數，似乎太多了，所以筆者介紹 LCD 2004
I²C 版(如下圖所示)。

(a). LCD2004 正面圖　　　(b). LCD 2004 I²C 轉接板　　　(c).LCD 2004 I²C 板

圖 42 LCD 2004I²C 板

　　如下圖所示，其實 LCD 2004 I²C 板是由標準 LCD 1602(如上圖.(a)所示)，加上
LCD 2004 I²C 轉接板轉接板(如上圖.(b)所示)，所組合出來的 LCD 2004 I²C 板(如上
圖.(c)所示)，讀者可以先買標準 LCD 1602(如上圖.(a)所示)，有需要的時後在買轉接

板(如上圖.(b)所示)，就可以組合成如下圖所示的成品。

圖 43 LCD 2004 I2C 零件表

圖 44 LCD 2004 I2C 組合圖

為了讓實驗更順暢進行，先參考下表所示之 LCD 2004 I2C 接腳表，將 LCD 2004 I2C 板與 Arduino 開發板進行實體線路連接，參考本文 LCD 2004 函式庫 單元，可以見到 LCD 2004 I2C 常用的函式庫 (LiquidCrystal Library, 參考網址：http://arduino.cc/en/Reference/LiquidCrystal ，http://playground.arduino.cc/Code/LCDi2c)。

表 19 LiquidCrystal Library API 相容表

Library	Displays Supported	Verified API	Connection
LCDi2cR	Robot-Electronics	Y	i2c
LCDi2cW	web4robot.com	Y	i2c
LiquidCrystal	Generic HitachiHD44780	P	4, 8 bit
LiquidCrystal_I2C	PCF8574drivingHD44780	Y	I2C

| LCDi2cNHD | NewHavenDisplayI2CMode | Y | i2c |
| ST7036 Lib | GenericST7036LCD controller | Y | i2c |

資料來源：Arduino 官網：http://playground.arduino.cc/Code/LCDAPI

由於不同種類的 Arduino 開發板，其 I2C/ TWI 接腳也略有不同，所以讀者可以參考下表所示之 Arduino 開發板 I2C/ TWI 接腳表，在根據下下下表之 LCD 1602 I2C 測試程式的內容，進行硬體接腳的修正，至於軟體部份，Arduino 軟體原始碼的部份，則不需要修正。

表 20 Arduino 開發板 I2C/ TWI 接腳表

開發板種類	I2C/ TWI 接腳表
Uno, Ethernet	A4 (SDA), A5 (SCL)
Mega2560	20 (SDA), 21 (SCL)
Leonardo	2 (SDA), 3 (SCL)
Due	20 (SDA), 21 (SCL),SDA1,SCL1

我們參考 Arduino 官方網站 http://arduino.cc/en/Reference/LiquidCrystal ，其連接 LCD 2004 範例程式，可以了解 Arduino 如何驅動 LCD 2004 顯示器：

表 21 LCD 2004 I2C 接腳表

接腳	接腳說明	接腳名稱
1	Ground (0V)	接地 (0V) Arduino GND
2	Supply voltage; 5V (4.7V － 5.3V)	電源 (+5V) Arduino +5V
3	SDA	Arduino digital Pin20(SDA)
4	SCL	Arduino digital Pin21(SCL)

接腳	接腳說明	接腳名稱

　　如下表所示，為 LCD 2004 I2C 測試程式，請讀者鍵入Ｓｋｅｔｃｈ　ＩＤＥ 軟體，編譯完成後上傳到開發版進行測試。

表 22 LCD 2004 I2C 測試程式

LCD 1602 I2C 測試程式(lcd2004_I2C_mill)

```
//Compatible with the Arduino IDE 1.0
//Library version:1.1
#include <Wire.h>
#include <LiquidCrystal_I2C.h>

LiquidCrystal_I2C lcd(0x27, 16, 2); // set the LCD address to 0x27 for a 16 chars and 2
line display

void setup()
{
  lcd.init();                         // initialize the lcd

  // Print a message to the LCD.
  lcd.backlight();
  lcd.print("Hello, world!");
}

void loop()
{
  // 將游標設到　 第一行,　 第二列
  // (注意:　 第二列第五行，因為是從 0 開始數起):
  lcd.setCursor(5, 1);
  // 列印 Arduino 重開之後經過的秒數
  lcd.print(millis() / 1000);
```

```
    Serial.println(millis() / 1000);
    delay(200);
}
```

讀者也可以在筆者 YouTube 頻道

(https://www.youtube.com/user/UltimaBruce)中，在網址

https://www.youtube.com/watch?v=GXAplXXnVn8&feature=youtu.be，看到本次實驗-

LCD 1602 I2C 測試程式結果畫面。

當然、如下圖所示，我們可以看到 Arduino 在 LCD 2004 畫面上顯示文字情形。

圖 45 LCD 1602 I2C 測試程式結果畫面

LCD 函數用法

為了更能了解 LCD 1602/2004 的用法，本節詳細介紹了 LiquidCrystal 函式主要
的用法：

LiquidCrystal(rs, enable, d0, d1, d2, d3, d4, d5, d6, d7)

1. 指令格式 LiquidCrystal lcd 物件名稱(使用參數)

2. 使用參數個格式如下：

LiquidCrystal(rs, enable, d4, d5, d6, d7)

LiquidCrystal(rs, enable, d0, d1, d2, d3,d4, d5, d6, d7)

LiquidCrystal(rs, rw, enable, d4, d5, d6, d7)

LiquidCrystal(rs, rw, enable, d0, d1, d2, d3, d4, d5, d6, d7)

LiquidCrystal.begin(16, 2)

1. 規劃 lcd 畫面大小(行寬，列寬)

2. 指令範例：

   LiquidCrystal.begin(16, 2)

   解釋：將目前 lcd 畫面大小，設成二列 16 行

LiquidCrystal.setCursor(0, 1)

1. LiquidCrystal.setCursor(行位置,列位置)，行位置從0開始,列位置從0開始(Arduino 第一都是從零開始)

2. 指令範例：

   LiquidCrystal.setCursor(0, 1)

   解釋：將目前游標跳到第一列第一行,為兩列,每列有 16 個字元(Arduino 第一都是從零開始)

LiquidCrystal.print()

1. LiquidCrystal.print (資料)，資料可以為 char, byte, int, long, or string

2. 指令範例：

   lcd.print("hello, world!");

   解釋：將目前游標位置印出『hello, world!』

LiquidCrystal.autoscroll()

1. 將目前 lcd 列印資料形態，設成可以捲軸螢幕

2. 指令範例：

   lcd.autoscroll();

   解釋：如使用 lcd.print(thisChar);，會將字元輸出到目前行列的位置，每輸出一個字元，行位置則加一，到第 16 字元時，若仍繼續輸出，則原有的列內的資料自動依 LiquidCrystal - Text Direction 的設定進行捲動，讓 print() 的命令繼續印出下個字元

LiquidCrystal.noAutoscroll()

1. 將目前 lcd 列印資料形態，設成不可以捲軸螢幕

2. 指令範例：

lcd.noAutoscroll();

解釋：如使用 lcd.print(thisChar); ，會將字元輸出到目前行列的位置，每輸出一個字元，行位置則加一，到第 16 字元時，若仍繼續輸出，讓 print() 的因繼續印出下個字元到下一個位置，但位置已經超越 16 行，所以輸出字元看不見。

LiquidCrystal.blink()

1. 將目前 lcd 游標設成閃爍

2. 指令範例：

lcd.blink();

解釋：將目前 lcd 游標設成閃爍

LiquidCrystal.noBlink()

1. 將目前 lcd 游標設成不閃爍

2. 指令範例：

lcd.noBlink ();

解釋：將目前 lcd 游標設成不閃爍

LiquidCrystal.cursor()

1. 將目前 lcd 游標設成底線狀態

2. 指令範例：

lcd.cursor();

解釋：將目前 lcd 游標設成底線狀態

LiquidCrystal.clear()

1. 將目前 lcd 畫面清除，並將游標位置回到左上角

2. 指令範例：

lcd.clear();

解釋：將目前 lcd 畫面清除，並將游標位置回到左上角

LiquidCrystal.home()

1. 將目前 lcd 游標位置回到左上角

2. 指令範例：

lcd.home();

解釋：將目前 lcd 游標位置回到左上角

章節小結

本章節介紹 LCD 顯示螢幕，主要是讓讀者了解 Arduino 開發板如何顯示資訊到外界的顯示裝置，透過以上章節的內容，一定可以一步一步的將資訊顯示給予實作出來。

5

CHAPTER

七段顯示器顯示模組

顯示七段顯示器

七段顯示器是 Arduino 開發板最常使用的數字顯示器，本實驗仍只需要一塊 Arduino 開發板、USB 下載線、單字元的七段顯示器。

如下圖所示，這個實驗我們需要用到的實驗硬體有下圖.(a)的 Arduino Mega 2560 與圖 46.(b) USB 下載線、下圖.(c)單位數七段顯示器：(曹永忠, 許碩芳, et al., 2015a; 曹永忠, 許碩芳, 許智誠, & 蔡英德, 2015b)

(a).Arduino Mega 2560

(b). USB 下載線

(c). 單位數七段顯示器

圖 46 顯示七段顯示器所需材料表

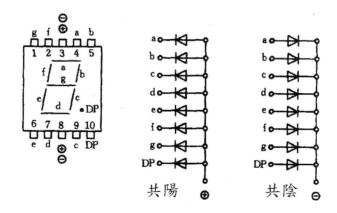

圖 47 七段顯示器接腳說明

我們遵照前幾章所述，將 Arduino 開發板的驅動程式安裝好之後，筆者參考上圖來了解單位數的七段顯示器接腳說明後，本實驗使用共陽型七段顯示器，所以筆者參考下表完成電路的連接，完成後如下圖所示。

表 23　七段顯示器接腳表

七段顯示器	Arduino 開發板接腳	解說
共陽(共陰)	Arduino Pin 5V	5V 陽極接點
七段顯示器.a	Arduino Pin 22	顯示 a 字形
七段顯示器.b	Arduino Pin 24	顯示 b 字形
七段顯示器.c	Arduino Pin 26	顯示 c 字形
七段顯示器.d	Arduino Pin 28	顯示 d 字形
七段顯示器.e	Arduino Pin 30	顯示 e 字形
七段顯示器.f	Arduino Pin 32	顯示 f 字形
七段顯示器.g	Arduino Pin 34	顯示 g 字形
七段顯示器.dot	Arduino Pin 36	顯示點 字形

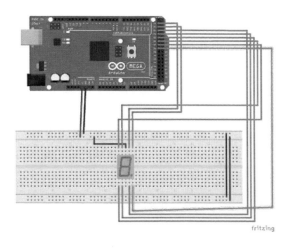

圖 48 單位七段顯示器接腳完成圖

我們遵照前幾章所述，將 Arduino 開發板的驅動程式安裝好之後，我們打開 Arduino 開發板的開發工具：Sketch IDE 整合開發軟體，攥寫一段程式，如下表所示之顯示七段顯示器測試程式，可以看到七段顯示器從 9、8、7、6、5、4、3、2、1、0 循環顯示。

表 24 顯示七段顯示器測試程式

顯示七段顯示器測試程式(7Segment)
// 七段顯示器製作倒數功能 (vturnon)
#define aPin 22
#define bPin 24
#define cPin 26
#define dPin 28
#define ePin 30
#define fPin 32
#define gPin 34
#define dotPin 36
#define turnon LOW
#define turnoff HIGH
void setup() {
PinMode(aPin, OUTPUT);
PinMode(bPin, OUTPUT);
PinMode(cPin, OUTPUT);
PinMode(dPin, OUTPUT);
PinMode(ePin, OUTPUT);
PinMode(fPin, OUTPUT);
PinMode(gPin, OUTPUT);
PinMode(dotPin, OUTPUT);
digitalWrite(dotPin, turnoff);   // 關閉小數點
}
void loop() {
// 顯示數字 '9'
digitalWrite(aPin, turnon);
digitalWrite(bPin, turnon);

```
digitalWrite(cPin, turnon);
digitalWrite(dPin, turnoff);
digitalWrite(ePin, turnoff);
digitalWrite(fPin, turnon);
digitalWrite(gPin, turnon);
delay(1000);
// 顯示數字 '8'
digitalWrite(aPin, turnon);
digitalWrite(bPin, turnon);
digitalWrite(cPin, turnon);
digitalWrite(dPin, turnon);
digitalWrite(ePin, turnon);
digitalWrite(fPin, turnon);
digitalWrite(gPin, turnon);
delay(1000);
// 顯示數字 '7'
digitalWrite(aPin, turnon);
digitalWrite(bPin, turnon);
digitalWrite(cPin, turnon);
digitalWrite(dPin, turnoff);
digitalWrite(ePin, turnoff);
digitalWrite(fPin, turnoff);
digitalWrite(gPin, turnoff);
delay(1000);
// 顯示數字 '6'
digitalWrite(aPin, turnon);
digitalWrite(bPin, turnoff);
digitalWrite(cPin, turnon);
digitalWrite(dPin, turnon);
digitalWrite(ePin, turnon);
digitalWrite(fPin, turnon);
digitalWrite(gPin, turnon);
delay(1000);
// 顯示數字 '5'
digitalWrite(aPin, turnon);
digitalWrite(bPin, turnoff);
digitalWrite(cPin, turnon);
digitalWrite(dPin, turnon);
digitalWrite(ePin, turnoff);
```

```
digitalWrite(fPin, turnon);
digitalWrite(gPin, turnon);
delay(1000);
// 顯示數字 '4'
digitalWrite(aPin, turnoff);
digitalWrite(bPin, turnon);
digitalWrite(cPin, turnon);
digitalWrite(dPin, turnoff);
digitalWrite(ePin, turnoff);
digitalWrite(fPin, turnon);
digitalWrite(gPin, turnon);
delay(1000);
// 顯示數字 '3'
digitalWrite(aPin, turnon);
digitalWrite(bPin, turnon);
digitalWrite(cPin, turnon);
digitalWrite(dPin, turnon);
digitalWrite(ePin, turnoff);
digitalWrite(fPin, turnoff);
digitalWrite(gPin, turnon);
delay(1000);
// 顯示數字 '2'
digitalWrite(aPin, turnon);
digitalWrite(bPin, turnon);
digitalWrite(cPin, turnoff);
digitalWrite(dPin, turnon);
digitalWrite(ePin, turnon);
digitalWrite(fPin, turnoff);
digitalWrite(gPin, turnon);
delay(1000);
// 顯示數字 '1'
digitalWrite(aPin, turnoff);
digitalWrite(bPin, turnon);
digitalWrite(cPin, turnon);
digitalWrite(dPin, turnoff);
digitalWrite(ePin, turnoff);
digitalWrite(fPin, turnoff);
digitalWrite(gPin, turnoff);
delay(1000);
```

```
// 顯示數字 '0'
digitalWrite(aPin, turnon);
digitalWrite(bPin, turnon);
digitalWrite(cPin, turnon);
digitalWrite(dPin, turnon);
digitalWrite(ePin, turnon);
digitalWrite(fPin, turnon);
digitalWrite(gPin, turnoff);
// 暫停 4 秒鐘
delay(4000);
}
```

資料來源：coopermaa(http://coopermaa2nd.blogspot.tw/2010/12/arduino-lab7.html)

讀者也可以在筆者 YouTube 頻道(https://www.youtube.com/user/UltimaBruce)中，
在網址 https://www.youtube.com/watch?v=YxMZvS8LWHA&feature=youtu.be ，看到本
次實驗-顯示七段顯示器測試程式，可以看到七段顯示器從 9、8、7、6、5、4、3、
2、1、0 循環顯示。

當然、如下圖所示， Arduino 開發板可以顯示七段顯示器測試程式，可以看到
七段顯示器從 9、8、7、6、5、4、3、2、1、0 循環顯示。

圖 49 顯示七段顯示器結果畫面

顯示二位數七段顯示器

二位數七段顯示器是 Arduino 開發板最常使用的數字顯示器，本實驗仍只需要一塊 Arduino 開發板、USB 下載線、二位數七段顯示器。

如下圖所示，這個實驗我們需要用到的實驗硬體有下圖.(a)的 Arduino Mega 2560 與下圖.(b) USB 下載線、下圖.(c) 二位數七段顯示器：

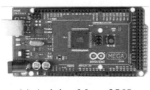

(a).Arduino Mega 2560

(b). USB 下載線

(c). 二位數七段顯示器

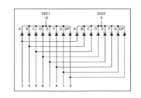

(d). 二位數七段顯示器接腳

圖 50 二位數七段顯示器所需材料表

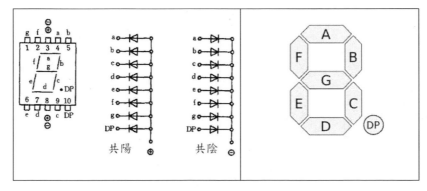

圖 51 七段顯示器接腳說明

我們遵照前幾章所述，將 Arduino 開發板的驅動程式安裝好之後，筆者參考上上圖、上圖來了解二位七段顯示器接腳說明後，本實驗使用共陽型二位七段顯示器，所以筆者參考下表所示之二位七段顯示器接腳表完成電路的連接，完成後如下圖所示。

表 25　二位七段顯示器接腳表

二位七段顯示器	Arduino 開發板接腳	解說
第一位數(共陽)(1)	Arduino Pin 7	控制第一位數顯示
第二位數(共陽)(10)	Arduino Pin 8	控制第二位數顯示
七段顯示器.a(4)	Arduino Pin 22	顯示 a 字形
七段顯示器.b(5)	Arduino Pin 24	顯示 b 字形
七段顯示器.c(9)	Arduino Pin 26	顯示 c 字形
七段顯示器.d(6)	Arduino Pin 28	顯示 d 字形
七段顯示器.e(8)	Arduino Pin 30	顯示 e 字形
七段顯示器.f(3)	Arduino Pin 32	顯示 f 字形
七段顯示器.g(2)	Arduino Pin 34	顯示 g 字形
七段顯示器.dot(7)	Arduino Pin 36	顯示點 字形

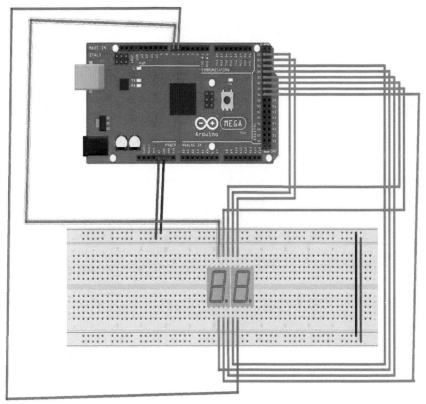

fritzing

圖 52 二位數七段顯示器完成圖

我們遵照前幾章所述，將 Arduino 開發板的驅動程式安裝好之後，我們打開 Arduino 開發板的開發工具：Sketch IDE 整合開發軟體，撰寫一段程式，如下表所示之二位數七段顯示器試程式，可以看到二位數七段顯示器從 99、98、97、96、95、......、3、2、1、0 循環顯示。

表 26 二位數七段顯示器測試程式

二位數七段顯示器測試程式(2DSegment)
// 七段顯示器製作倒數功能（vturnon）
#define ctlD1 7

```
#define ctlD2 6

#define aPin 22
#define bPin 24
#define cPin 26
#define dPin 28
#define ePin 30
#define fPin 32
#define gPin 34
#define dotPin 36
#define turnon LOW
#define turnoff HIGH
#define digitalon HIGH
#define digitaloff LOW

void setup() {
    PinMode(ctlD1, OUTPUT);
    PinMode(ctlD2, OUTPUT);
    PinMode(aPin, OUTPUT);
    PinMode(bPin, OUTPUT);
    PinMode(cPin, OUTPUT);
    PinMode(dPin, OUTPUT);
    PinMode(ePin, OUTPUT);
    PinMode(fPin, OUTPUT);
    PinMode(gPin, OUTPUT);
    PinMode(dotPin, OUTPUT);
    digitalWrite(dotPin, turnoff);    // 關閉小數點
}

void loop() {
      int i ;
    for (i=0 ; i<99; i++)
            {
             showNumber(i) ;
            delay(50) ;
            showNumber(-1) ;

        }
}
```

```
void showNumber(int no)
{
    if (no == -1)
    {
        ShowSegment(1, -1) ;
    }
    else
    {
        ShowSegment(1, (no/10)) ;
        delayMicroseconds(3000) ;
        ShowSegment(2, (no%10)) ;
        delayMicroseconds(3000) ;
    }
}
void ShowSegment(int digital, int number)
{
    if (digital == 1)
        {
            digitalWrite(ctlD1, digitalon);
            digitalWrite(ctlD2, digitaloff);
        }
        else
        {
            digitalWrite(ctlD1, digitaloff);
            digitalWrite(ctlD2, digitalon);
        }


    switch (number)
        {
            case 9:
                    // 顯示數字 '9'
                digitalWrite(aPin, turnon);
                digitalWrite(bPin, turnon);
                digitalWrite(cPin, turnon);
                digitalWrite(dPin, turnoff);
                digitalWrite(ePin, turnoff);
                digitalWrite(fPin, turnon);
```

```
digitalWrite(gPin, turnon);
 break ;

 case 8:
// 顯示數字 '8'
digitalWrite(aPin, turnon);
digitalWrite(bPin, turnon);
digitalWrite(cPin, turnon);
digitalWrite(dPin, turnon);
digitalWrite(ePin, turnon);
digitalWrite(fPin, turnon);
digitalWrite(gPin, turnon);
break ;

case 7:
// 顯示數字 '7'
digitalWrite(aPin, turnon);
digitalWrite(bPin, turnon);
digitalWrite(cPin, turnon);
digitalWrite(dPin, turnoff);
digitalWrite(ePin, turnoff);
digitalWrite(fPin, turnoff);
digitalWrite(gPin, turnoff);
break ;

case 6:
// 顯示數字 '6'
digitalWrite(aPin, turnon);
digitalWrite(bPin, turnoff);
digitalWrite(cPin, turnon);
digitalWrite(dPin, turnon);
digitalWrite(ePin, turnon);
digitalWrite(fPin, turnon);
digitalWrite(gPin, turnon);
break ;

case 5:
// 顯示數字 '5'
digitalWrite(aPin, turnon);
```

```
digitalWrite(bPin, turnoff);
digitalWrite(cPin, turnon);
digitalWrite(dPin, turnon);
digitalWrite(ePin, turnoff);
digitalWrite(fPin, turnon);
digitalWrite(gPin, turnon);
break ;

case 4:
// 顯示數字 '4'
digitalWrite(aPin, turnoff);
digitalWrite(bPin, turnon);
digitalWrite(cPin, turnon);
digitalWrite(dPin, turnoff);
digitalWrite(ePin, turnoff);
digitalWrite(fPin, turnon);
digitalWrite(gPin, turnon);
break ;

case 3:
// 顯示數字 '3'
digitalWrite(aPin, turnon);
digitalWrite(bPin, turnon);
digitalWrite(cPin, turnon);
digitalWrite(dPin, turnon);
digitalWrite(ePin, turnoff);
digitalWrite(fPin, turnoff);
digitalWrite(gPin, turnon);
break ;

 case 2:
// 顯示數字 '2'
digitalWrite(aPin, turnon);
digitalWrite(bPin, turnon);
digitalWrite(cPin, turnoff);
digitalWrite(dPin, turnon);
digitalWrite(ePin, turnon);
digitalWrite(fPin, turnoff);
digitalWrite(gPin, turnon);
```

```
        break ;

        case 1:
        // 顯示數字 '1'
        digitalWrite(aPin, turnoff);
        digitalWrite(bPin, turnon);
        digitalWrite(cPin, turnon);
        digitalWrite(dPin, turnoff);
        digitalWrite(ePin, turnoff);
        digitalWrite(fPin, turnoff);
        digitalWrite(gPin, turnoff);
        break ;

        case 0:
        // 顯示數字 '0'
        digitalWrite(aPin, turnon);
        digitalWrite(bPin, turnon);
        digitalWrite(cPin, turnon);
        digitalWrite(dPin, turnon);
        digitalWrite(ePin, turnon);
        digitalWrite(fPin, turnon);
        digitalWrite(gPin, turnoff);
          break ;

        case -1:
        // all Off
        digitalWrite(aPin, turnoff);
        digitalWrite(bPin, turnoff);
        digitalWrite(cPin, turnoff);
        digitalWrite(dPin, turnoff);
        digitalWrite(ePin, turnoff);
        digitalWrite(fPin, turnoff);
        digitalWrite(gPin, turnoff);
          break ;

    }
}
```

資料來源：Modifed from coopermaa(http://coopermaa2nd.blogspot.tw/2010/12/arduino-lab7.html)

讀者也可以在筆者 YouTube 頻道(https://www.youtube.com/user/UltimaBruce)中，在網址 https://www.youtube.com/watch?v=Gu81t8U8XNA&feature=youtu.be ，看到本次實驗-顯示二位數七段顯示器測試程式，可以看到二位數七段顯示器從 99、98、97、96、95、......、3、2、1、0 循環顯示。

當然、如下圖所示，Arduino 開發板可以顯示二位數七段顯示器測試程式，可以看到七段顯示器從 99、98、97、96、95、......、3、2、1、0 循環顯示。

圖 53 二位數七段顯示器結果畫面

顯示四位數七段顯示器

四位七段顯示器是 Arduino 開發板最常使用的多數字顯示器，本實驗仍只需要一塊 Arduino 開發板、USB 下載線、四位數七段顯示器。

如圖 54 下圖所示，這個實驗我們需要用到的實驗硬體有下圖.(a)的 Arduino Mega 2560 與下圖.(b) USB 下載線、下圖.(c) 四位數七段顯示器：

(a).Arduino Mega 2560

(b). USB 下載線

(c).四位數七段顯示器

(d). 四位數七段顯示器接腳圖

圖 54 四位數七段顯示器所需零件表

圖 55 四位數七段顯示器接腳說明

　　我們遵照前幾章所述，將 Arduino 開發板的驅動程式安裝好之後，筆者參考上

上圖、上圖來了解四位數七段顯示器接腳說明後，本實驗使用共陰型四位數七段顯

示器，所以筆者參考下表完成電路的連接，完成後如下圖所示。。

表 27　四位數七段顯示器接腳表

七段顯示器	Arduino 開發板接腳	解說
第四位數字控制 (共陰)	Arduino Pin 7	第四位數字控制
第三位數字控制 (共陰)	Arduino Pin 6	第三位數字控制
第二位數字控制 (共陰)	Arduino Pin 5	第二位數字控制
第一位數字控制 (共陰)	Arduino Pin 4	第一位數字控制

七段顯示器	Arduino 開發板接腳	解說
七段顯示器.a	Arduino Pin 22	顯示 a 字形
七段顯示器.b	Arduino Pin 24	顯示 b 字形
七段顯示器.c	Arduino Pin 26	顯示 c 字形
七段顯示器.d	Arduino Pin 28	顯示 d 字形
七段顯示器.e	Arduino Pin 30	顯示 e 字形
七段顯示器.f	Arduino Pin 32	顯示 f 字形
七段顯示器.g	Arduino Pin 34	顯示 g 字形
七段顯示器.dot	Arduino Pin 36	顯示點 字形

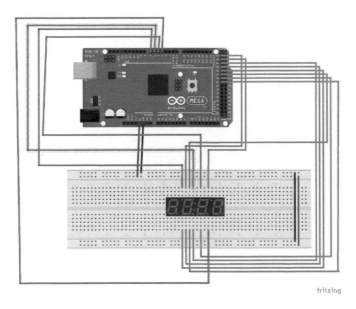

圖 56 四位數七段顯示器接腳完成圖

我們遵照前幾章所述,將 Arduino 開發板的驅動程式安裝好之後,我們打開 Arduino 開發板的開發工具:Sketch IDE 整合開發軟體,攥寫一段程式,如表 28 下表所示之顯示四位數七段顯示器測試程式,可以看到四位數七段顯示器從 0001、

0002、....、9998、9999 循環顯示。

表 28 四位數七段顯示器測試程式

四位數七段顯示器測試程式(4DSegment)

// 七段顯示器製作倒數功能（vturnon）
#define ctlD1 7
#define ctlD2 6
#define ctlD3 5
#define ctlD4 4
 	
#define aPin 22
#define bPin 24
#define cPin 26
#define dPin 28
#define ePin 30
#define fPin 32
#define gPin 34
#define dotPin 36
#define turnon HIGH
#define turnoff LOW
#define digitalon LOW
#define digitaloff HIGH
int number = 0;
 	
unsigned long time_previous;
void setup() {
  PinMode(ctlD1, OUTPUT);
  PinMode(ctlD2, OUTPUT);
  PinMode(ctlD3, OUTPUT);
  PinMode(ctlD4, OUTPUT);
  PinMode(aPin, OUTPUT);
  PinMode(bPin, OUTPUT);
  PinMode(cPin, OUTPUT);
  PinMode(dPin, OUTPUT);
  PinMode(ePin, OUTPUT);
  PinMode(fPin, OUTPUT);
  PinMode(gPin, OUTPUT);

```
    PinMode(dotPin, OUTPUT);
    digitalWrite(dotPin, turnoff);    // 關閉小數點
     Serial.begin(9600);

}

void loop() {
      int i ;
   // 經過一秒後就讓 number 加 1
   unsigned long time_now = millis();
   if(time_now - time_previous > 1000){
      number++;
      time_previous += 1000;
      Serial.println("number=%d\n", number);
   }

   // 不斷地寫入數字
   showNumber(number);
}

 void showNumber(int no)
 {
    if (no == -1)
    {
         ShowSegment(1, -1) ;
    }
    else
    {
       ShowSegment(1, (no/1000)) ;
       delay(5) ;
       ShowSegment(2, (no/100)) ;
       delay(5) ;
       ShowSegment(3, (no/10)) ;
     delay(5) ;
       ShowSegment(4, (no%10)) ;
       delay(5) ;
    }
 }
 void ShowSegment(int digital, int number)
```

```
{
    switch (digital)
        {
            case 1:
            digitalWrite(ctlD1, digitalon);
            digitalWrite(ctlD2, digitaloff);
            digitalWrite(ctlD3, digitaloff);
            digitalWrite(ctlD4, digitaloff);
            break ;

            case 2:
            digitalWrite(ctlD1, digitaloff);
            digitalWrite(ctlD2, digitalon);
            digitalWrite(ctlD3, digitaloff);
            digitalWrite(ctlD4, digitaloff);
            break ;

            case 3:
            digitalWrite(ctlD1, digitaloff);
            digitalWrite(ctlD2, digitaloff);
            digitalWrite(ctlD3, digitalon);
            digitalWrite(ctlD4, digitaloff);
            break ;

            case 4:
            digitalWrite(ctlD1, digitaloff);
            digitalWrite(ctlD2, digitaloff);
            digitalWrite(ctlD3, digitaloff);
            digitalWrite(ctlD4, digitalon);
            break ;
        }

    switch (number)
    {
        case 9:
                    // 顯示數字 '9'
            digitalWrite(aPin, turnon);
            digitalWrite(bPin, turnon);
            digitalWrite(cPin, turnon);
```

```
digitalWrite(dPin, turnoff);
digitalWrite(ePin, turnoff);
digitalWrite(fPin, turnon);
digitalWrite(gPin, turnon);
 break ;

 case 8:
// 顯示數字 '8'
digitalWrite(aPin, turnon);
digitalWrite(bPin, turnon);
digitalWrite(cPin, turnon);
digitalWrite(dPin, turnon);
digitalWrite(ePin, turnon);
digitalWrite(fPin, turnon);
digitalWrite(gPin, turnon);
break ;

case 7:
// 顯示數字 '7'
digitalWrite(aPin, turnon);
digitalWrite(bPin, turnon);
digitalWrite(cPin, turnon);
digitalWrite(dPin, turnoff);
digitalWrite(ePin, turnoff);
digitalWrite(fPin, turnoff);
digitalWrite(gPin, turnoff);
break ;

case 6:
// 顯示數字 '6'
digitalWrite(aPin, turnon);
digitalWrite(bPin, turnoff);
digitalWrite(cPin, turnon);
digitalWrite(dPin, turnon);
digitalWrite(ePin, turnon);
digitalWrite(fPin, turnon);
digitalWrite(gPin, turnon);
break ;
```

```
case 5:
// 顯示數字 '5'
digitalWrite(aPin, turnon);
digitalWrite(bPin, turnoff);
digitalWrite(cPin, turnon);
digitalWrite(dPin, turnon);
digitalWrite(ePin, turnoff);
digitalWrite(fPin, turnon);
digitalWrite(gPin, turnon);
break ;

case 4:
// 顯示數字 '4'
digitalWrite(aPin, turnoff);
digitalWrite(bPin, turnon);
digitalWrite(cPin, turnon);
digitalWrite(dPin, turnoff);
digitalWrite(ePin, turnoff);
digitalWrite(fPin, turnon);
digitalWrite(gPin, turnon);
break ;

case 3:
// 顯示數字 '3'
digitalWrite(aPin, turnon);
digitalWrite(bPin, turnon);
digitalWrite(cPin, turnon);
digitalWrite(dPin, turnon);
digitalWrite(ePin, turnoff);
digitalWrite(fPin, turnoff);
digitalWrite(gPin, turnon);
break ;

 case 2:
// 顯示數字 '2'
digitalWrite(aPin, turnon);
digitalWrite(bPin, turnon);
digitalWrite(cPin, turnoff);
digitalWrite(dPin, turnon);
```

```
            digitalWrite(ePin, turnon);
            digitalWrite(fPin, turnoff);
            digitalWrite(gPin, turnon);
            break ;

        case 1:
        // 顯示數字 '1'
            digitalWrite(aPin, turnoff);
            digitalWrite(bPin, turnon);
            digitalWrite(cPin, turnon);
            digitalWrite(dPin, turnoff);
            digitalWrite(ePin, turnoff);
            digitalWrite(fPin, turnoff);
            digitalWrite(gPin, turnoff);
            break ;

        case 0:
        // 顯示數字 '0'
            digitalWrite(aPin, turnon);
            digitalWrite(bPin, turnon);
            digitalWrite(cPin, turnon);
            digitalWrite(dPin, turnon);
            digitalWrite(ePin, turnon);
            digitalWrite(fPin, turnon);
            digitalWrite(gPin, turnoff);
            break ;

        case -1:
        // all Off
            digitalWrite(aPin, turnoff);
            digitalWrite(bPin, turnoff);
            digitalWrite(cPin, turnoff);
            digitalWrite(dPin, turnoff);
            digitalWrite(ePin, turnoff);
            digitalWrite(fPin, turnoff);
            digitalWrite(gPin, turnoff);
            break ;

}
```

}

參考資料來源：coopermaa(http://yehnan.blogspot.tw/2013/08/arduino_26.html)

讀者也可以在筆者 YouTube 頻道(https://www.youtube.com/user/UltimaBruce)中，在網址 https://www.youtube.com/watch?v=DDF_y0BumG0&feature=youtu.be ，看到本次實驗-顯示四位數七段顯示器測試程式，如下圖所示，可以看到四位數七段顯示器從 0001、0002、....、9998、9999 循環顯示。

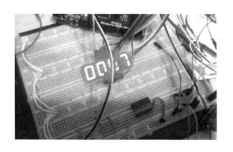

圖 57 顯示四位數七段顯示器結果畫面

章節小結

本章節介紹七段顯示器顯示模組，主要是讓讀者了解 Arduino 開發板如何顯示資訊到外界的顯示裝置，透過以上章節的內容，一定可以一步一步的將資訊顯示給予實作出來。

CHAPTER

Led 發光二極體顯示螢幕

本文實驗為了讓讀者可以更簡單驅動 Led 來顯示資訊內容，我們直覺可以使用基本的 Led 發光二極體，由下圖所示，將之並列成排，由排成陣，組合成一個完整的 Led 陣列，或許讀者覺的下圖所示的 Led 沒有太複雜，由下圖.(c)所示，共有 8*18 Led 排在一起，背後須要焊接 8*18 顆*2 針腳=>288 個焊接點，相信許多讀者看到就頭暈了，所以目前使用基本的 Led 發光二極體來排列成陣列的方式並不多見，因施工成本高、錯誤率高、維護成本高、易損壞、耗電量高等等眾多因素，所以越來越多人採用如圖 69 所示的矩陣式 LED(曹永忠, 許智誠, & 蔡英德, 2014b; 曹永忠, 許智誠, & 蔡英德, 2014b, 2014c, 2014d, 2014e, 2014f, 2015e, 2015k)。

(a).排在一起 Led　　　　(b).6*6 Led 陣列　　　　(c).8*18 Led 陣列

圖 58 Led 發光二極體組合的 Led 陣列

首先，本實驗使用上圖方式，由下圖方式，自行組成一個 7*5 LED 矩陣，由於我們必需將列接腳與行接腳，如下表所示，完成 Arduino 開發板與 7*5 矩陣式 LED 連接之後，將下表所示之 7*5 矩陣式 LED 測試程式一鍵入 Arduino Sketch 之中，完成編譯後，上載到 Arduino 開發板進行測試，可以看到 7*5 矩陣式 LED 亮出畫面。

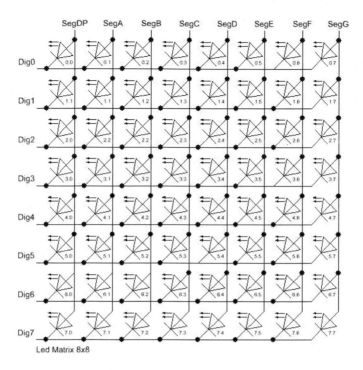

圖 59 8*8 Led 接腳範例圖

本實驗參考下表方式，進行電路組立。

表 29 7*5　LED 接腳表

	模組接腳	Arduino 開發板接腳	解說
Led Matrix 5*7	Row0	Arduino Pin 2	5 * 7 Led 列接腳
	Row1	Arduino Pin 3	
	Row2	Arduino Pin 4	
	Row3	Arduino Pin 5	
	Row4	Arduino Pin 6	
	Row5	Arduino Pin 7	
	Row6	Arduino Pin 8	
	Column0	Arduino Pin 13	5 * 7 Led 行接腳
	Column1	Arduino Pin 12	
	Column2	Arduino Pin 11	
	Column3	Arduino Pin 10	

模組接腳	Arduino 開發板接腳	解說
Column4	Arduino Pin 9	

我們，請讀者鍵入Ｓｋｅｔｃｈ　ＩＤＥ軟體(軟體下載請到：

https://www.arduino.cc/en/Main/Software)，編譯完成後上傳到開發版進行測試。

表 30 7*5 矩陣式 LED 測試程式一

```
7*5 矩陣式 LED 測試程式一(Led01)

/* Pin Mapping for Common Cathode 5x7 LED Matrix
    LED Matrix      Arduino(x = not connect)      Designator
        1               6                          Row4
        2               8                          Row6
        3               12                         Column1
        4               11                         Column2
        5               x
        6               9                          Column4
        7               7                          Row5
        8               4                          Row2
        9               2                          Row0
       10               10                         Column3
       11               x
       12               5                          Row3
       13               13                         Column0
       14               3                          Row1


    ===================================
    LED Matrix      Arduino(x = not connect)      Designator
        9               2                          Row0
       14               3                          Row1
        8               4                          Row2
       12               5                          Row3
        1               6                          Row4
        7               7                          Row5
        2               8                          Row6
       13               13                         Column0
```

3	12	Column1
4	11	Column2
10	10	Column3
6	9	Column4
5	x	
11	x	

```
==================================
    Arduino pin2~pin8 = row0~row6, so rowX +2 = pin number
    Arduino pin13~pin9 = column0~column4, 13 - columnY = pin number
*/

// 5x7 LED Matrix
const byte COLS = 5; // 5 Columns
const byte ROWS = 7; // 7 Rows

void clearLEDs() {
  for (int r = 0; r < ROWS; r++) {
    digitalWrite( r + 2, LOW);
  }

  for (int c = 0; c < COLS; c++) {
      digitalWrite( 13 - c, LOW);   // 共陰極，每一行預設均為接地(Low = 接
地，High = 熄燈)
  }
}

void setup() {
  // 5x7 LED Matrix 接在 pin 2 ~ pin 13
  // 把腳位設置為 output pin
  for (int i = 2; i <= 13; i++) {
    pinMode(i, OUTPUT);
  }
}

void loop() {
  clearLEDs();
```

7*5 矩陣式 LED 測試程式一(Led01)

```
// 由下而下，一排一排(Row)打開
for (int r = 0; r < ROWS; r++) {
    digitalWrite( r + 2, HIGH);
    delay(300);
}
/*
// 由左至右，一行一行(Column)關掉
for (int c = 0; c < COLS; c++) {
    digitalWrite(13 - c, HIGH);
    delay(300);
}
*/

// 暫停 1 秒鐘
delay(1000);
}
```

顯示 8x8 Led 點陣顯示器

Led 點陣顯示器是 Arduino 開發板最常使用來顯室圖形、英文字、甚至中文字的顯示器，本實驗仍只需要一塊 Arduino 開發板、USB 下載線、四位數七段顯示器。

如下圖所示，這個實驗我們需要用到的實驗硬體有下圖.(a)的 Arduino Mega 2560 下圖.(b) USB 下載線、下圖.(c) 8x8 Led 點陣顯示器：

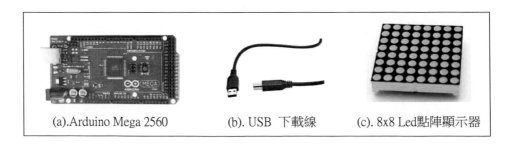

| (a).Arduino Mega 2560 | (b). USB 下載線 | (c). 8x8 Led點陣顯示器 |

圖 60 顯示 8x8 Led 點陣顯示器所需零件表

　　如下圖所示，我們可以知道 led 點陣 8x8 顯示器背面有接腳，可以見到『led 點陣 8x8 顯示器』下方有印字的為開始的腳位，由左到右共有 PIN1~PIN8，上面由右到左共有 PIN9~PIN16，讀者要仔細觀看，切勿弄混淆了。

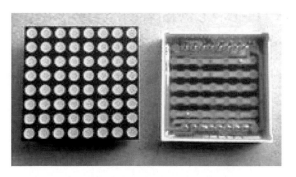

圖 61 led 點陣 8x8 顯示器接腳圖

　　如下圖所示，我們可以知道接腳的腳位，但是那一個腳位是代表那一個 LED 燈，我們可以由下圖得知規納成為下列表所示之 8x8 Led 點陣顯示器腳位、行列對照表，讀者可以透過此表得知如何驅動 8x8 Led 點陣顯示器。

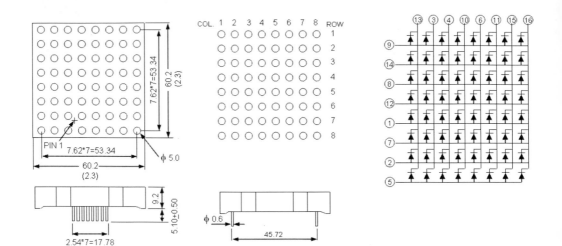

圖 62 8x8 Led 點陣顯示器接腳說明

表 31 8x8 Led 點陣顯示器腳位、行列對照表

8x8 Led 點陣顯示器	
8x8 Led 點陣顯示器腳位	驅動列行意義
8x8 Led 點陣顯示器 pin1	8x8 Led 點陣顯示器 R5
8x8 Led 點陣顯示器 pin2	8x8 Led 點陣顯示器 R7
8x8 Led 點陣顯示器 pin3	8x8 Led 點陣顯示器 C2
8x8 Led 點陣顯示器 pin4	8x8 Led 點陣顯示器 C3
8x8 Led 點陣顯示器 pin5	8x8 Led 點陣顯示器 R8
8x8 Led 點陣顯示器 pin6	8x8 Led 點陣顯示器 C5
8x8 Led 點陣顯示器 pin7	8x8 Led 點陣顯示器 R6
8x8 Led 點陣顯示器 pin8	8x8 Led 點陣顯示器 R3
8x8 Led 點陣顯示器 pin9	8x8 Led 點陣顯示器 R1
8x8 Led 點陣顯示器 pin10	8x8 Led 點陣顯示器 C4
8x8 Led 點陣顯示器 pin11	8x8 Led 點陣顯示器 C6
8x8 Led 點陣顯示器 pin12	8x8 Led 點陣顯示器 R4
8x8 Led 點陣顯示器 pin13	8x8 Led 點陣顯示器 C1
8x8 Led 點陣顯示器 pin14	8x8 Led 點陣顯示器 R2
8x8 Led 點陣顯示器 pin15	8x8 Led 點陣顯示器 C7
8x8 Led 點陣顯示器 pin16	8x8 Led 點陣顯示器 C8

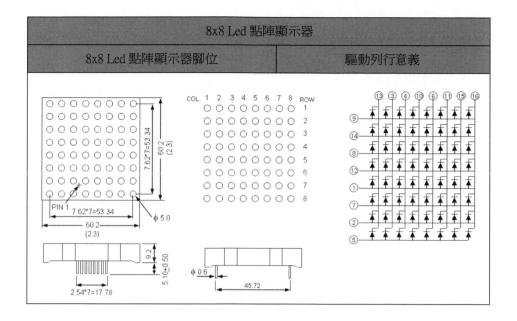

8x8 Led 點陣顯示器		
8x8 Led 點陣顯示器腳位		驅動列行意義

我們遵照前面所述,將 Arduino 開發板的驅動程式安裝好之後,筆者參考上表、上上圖、上圖了解之後,轉成下表所示之 8x8 Led 點陣顯示器接腳表,本實驗使用共陽型 8x8 Led 點陣顯示器接腳表,所以筆者參考下表完成電路的連接,完成後如下圖所示之 8x8 Led 點陣顯示器實際組裝圖。

表 32 8x8 Led 點陣顯示器接腳表

8x8 Led 點陣顯示器	Arduino 開發板接腳	解說
8x8 Led 點陣顯示器 pin1	Arduino pin 22	8x8 Led 點陣顯示器 R5
8x8 Led 點陣顯示器 pin2	Arduino pin 24	8x8 Led 點陣顯示器 R7
8x8 Led 點陣顯示器 pin3	Arduino pin 26	8x8 Led 點陣顯示器 C2
8x8 Led 點陣顯示器 pin4	Arduino pin 28	8x8 Led 點陣顯示器 C3
8x8 Led 點陣顯示器 pin5	Arduino pin 30	8x8 Led 點陣顯示器 R8
8x8 Led 點陣顯示器 pin6	Arduino pin 32	8x8 Led 點陣顯示器 C5
8x8 Led 點陣顯示器 pin7	Arduino pin 34	8x8 Led 點陣顯示器 R6
8x8 Led 點陣顯示器 pin8	Arduino pin 36	8x8 Led 點陣顯示器 R3
8x8 Led 點陣顯示器 pin9	Arduino pin 38	8x8 Led 點陣顯示器 R1
8x8 Led 點陣顯示器 pin10	Arduino pin 40	8x8 Led 點陣顯示器 C4
8x8 Led 點陣顯示器 pin11	Arduino pin 42	8x8 Led 點陣顯示器 C6

8x8 Led 點陣顯示器	Arduino 開發板按腳	解說
8x8 Led 點陣顯示器 pin12	Arduino pin 44	8x8 Led 點陣顯示器 R4
8x8 Led 點陣顯示器 pin13	Arduino pin 46	8x8 Led 點陣顯示器 C1
8x8 Led 點陣顯示器 pin14	Arduino pin 48	8x8 Led 點陣顯示器 R2
8x8 Led 點陣顯示器 pin15	Arduino pin 50	8x8 Led 點陣顯示器 C7
8x8 Led 點陣顯示器 pin16	Arduino pin 52	8x8 Led 點陣顯示器 C8

圖 63 8x8 Led 點陣顯示器實際組裝圖

我們遵照前幾章所述,將 Arduino 開發板的驅動程式安裝好之後,請讀者鍵入Ｓｋｅｔｃｈ ＩＤＥ軟體 (軟體下載請到:

https://www.arduino.cc/en/Main/Software),編譯完成後上傳到開發版進行測試。

如下表所示之 8x8 Led 點陣顯示器測試程式一，可以看到 8x8 Led 點陣顯示器
一列一列亮起來。

表 33 8x8 Led 點陣顯示器測試程式一

Led 點陣顯示器測試程式一(ledmatrix_LineUp)

```
#define turnon HIGH
#define turnoff LOW


byte rows[8] = {9, 14, 8, 12, 1, 7, 2, 5};
byte cols[8] = {13, 3, 4, 10, 6, 11, 15, 16};
//byte pins[16] = {5, 4, 3, 2, 14, 15, 16, 17, 13, 12, 11, 10, 9, 8, 7, 6};
byte pins[16] ={22,24, 26, 28, 30, 32,34,36,38,40,42,44,46,48,50,52};
byte screen[8] = {0, 0, 0, 0, 0, 0, 0, 0};

void setup() {
   for (int i = 0; i <= 15; i++)
       {
            pinMode(pins[i], OUTPUT);
            digitalWrite(pins[i],turnoff);
       }
       delay(1000);

       for (int i = 0; i <= 7; i++)
         {
              digitalWrite(pins[rows[i]-1],turnon);
              delay(1000);
         }


    //for (i=1;i<=8, i++)
}


void loop() {
 // digitalWrite(4,HIGH);
}
```

讀者也可以在筆者 YouTube 頻道(https://www.youtube.com/user/UltimaBruce)中，在網址：https://www.youtube.com/watch?v=JDgZxvtLkBY&feature=youtu.be，看到本次實驗- Led 點陣顯示器測試程式一結果畫面，如下圖所示，以看到 8x8 Led 點陣顯示器一列一列亮起來。

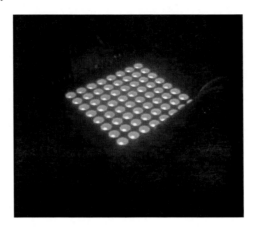

圖 64 Led 點陣顯示器測試程式一結果畫面

滑動顯示 8x8 Led 點陣顯示器

我們遵照前面文章所述，已將 8x8 Led 點陣顯示器顯示出來，但是只是點亮 8x8 Led 點陣顯示器，如果我們要顯示圖形或文字，由於『8x8 Led 點陣顯示器』需要許多時間顯示，在顯示 8x8 Led 點陣顯示器期間又不可以中斷，這樣會讓圖形不見或停窒，這是因為『8x8 Led 點陣顯示器』的內容是透過每一點點亮，熄滅，掃描每一點之後換列，在往下一列開始，如下圖所示，所以必須一點一點的點亮，由於人類視覺暫留的原理，人類的眼睛與頭腦會自動將之合成一幅完整的畫面(曹永忠, 許智誠, et al., 2014b; 曹永忠, 許智誠, et al., 2014b, 2014c, 2014d, 2014e, 2014f, 2015e, 2015k)。

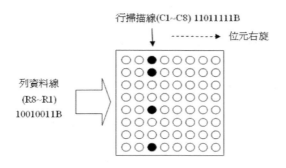

圖 65 Led 點陣顯示掃描方式

資料來源：亞洲大學資工系 陳瑞奇(Rikki Chen, CSIE, Asia
Univ.)(http://dns2.asia.edu.tw/~rikki/mps103/mcs-ch12.pdf)

由下圖所示之 Led 點陣顯示器掃描電路，由於 8 x 8 點的 Led 點陣顯示器，共
有 8*8 點，必須將之分成八列八行，將 8 個行接點(C1~C8)與 8 個列接點(R1~R8)，
規劃成 8 條資料線與 8 條掃描線。每次資料線送出 1 行編碼資料(1Bytes)，使用掃
描線，選擇其中一行輸出，經短暫時間，送出下一行編碼資料，8 行輪流顯示，利
用眼睛視覺暫留效應，看到整個編碼圖形的顯示(曹永忠, 許智誠, & 蔡英德, 2014a;
曹永忠, 許智誠, & 蔡英德, 2014a; 曹永忠, 許智誠, et al., 2014b, 2014c, 2014f)。

如果我們要顯示一個『u』字元在 Led 點陣顯示器上，如下表所示，我們必須
先將『u』字元的外形進行編碼，用 8 bit *8 列來進行編碼。

表 34 顯示 u 編碼表

掃瞄順序	顯示資料（2 進位制）	顯示資料（16 進位制）
第 1 行	00001000	0x08
第 2 行	00100100	0x24
第 3 行	01010010	0x52
第 4 行	01001000	0x48
第 5 行	01000001	0x41
第 6 行	00100010	0x22
第 7 行	01000100	0x44
第 8 行	00001000	0x08

然後再將上表所示之 8bits *8，如下圖所示，一列一點的顯示出來。

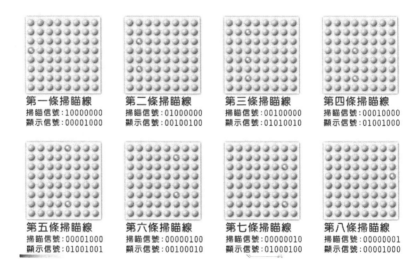

第一條掃瞄線
掃瞄信號：10000000
顯示信號：00001000

第二條掃瞄線
掃瞄信號：01000000
顯示信號：00100100

第三條掃瞄線
掃瞄信號：00100000
顯示信號：01010010

第四條掃瞄線
掃瞄信號：00010000
顯示信號：01001000

第五條掃瞄線
掃瞄信號：00001000
顯示信號：01001001

第六條掃瞄線
掃瞄信號：00000100
顯示信號：00100010

第七條掃瞄線
掃瞄信號：00000010
顯示信號：01000100

第八條掃瞄線
掃瞄信號：00000001
顯示信號：00001000

圖 66 顯示 u 之掃描方法

如此一來，Arduino 開發板就必須不中斷的顯示這些 Led 點陣顯示器的所有點，這樣 Arduino 開發板跟本就無力去做別的事，所以我們引入了 Timer 的方法，來重新改寫這個程式。

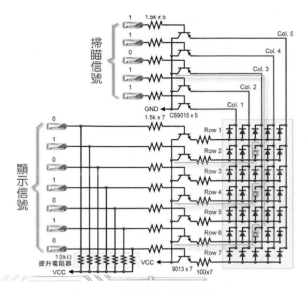

圖 67 Led 點陣顯示掃描電路

資料來源：亞洲大學資工系 陳瑞奇(Rikki Chen, CSIE, Asia Univ.)(http://dns2.asia.edu.tw/~rikki/mps103/mcs-ch12.pdf)

在 Arduino 開發板的驅動程式安裝好之後，我們，請讀者鍵入Ｓｋｅｔｃｈ ＩＤＥ軟體(軟體下載請到：https://www.arduino.cc/en/Main/Software)，編譯完成後上傳到開發版進行測試。

如表所示之滑動顯示 8x8 Led 點陣顯示器測試程式，可以看到 8x8 Led 點陣顯示器滑動顯示『HELLO』。

表 35 滑動顯示 8x8 Led 點陣顯示器測試程式

滑動顯示 8x8 Led 點陣顯示器測試程式(ledmatrix_HELLO)
#include <FrequencyTimer2.h>   #define turnon HIGH   #define turnoff LOW   #define SPACE { \   {0, 0, 0, 0, 0, 0, 0, 0},   \

```
        {0, 0, 0, 0, 0, 0, 0, 0}, \
        {0, 0, 0, 0, 0, 0, 0, 0}, \
        {0, 0, 0, 0, 0, 0, 0, 0}, \
        {0, 0, 0, 0, 0, 0, 0, 0}, \
        {0, 0, 0, 0, 0, 0, 0, 0}, \
        {0, 0, 0, 0, 0, 0, 0, 0}, \
        {0, 0, 0, 0, 0, 0, 0, 0} \
}

#define H { \
        {0, 1, 0, 0, 0, 0, 1, 0}, \
        {0, 1, 0, 0, 0, 0, 1, 0}, \
        {0, 1, 0, 0, 0, 0, 1, 0}, \
        {0, 1, 1, 1, 1, 1, 1, 0}, \
        {0, 1, 0, 0, 0, 0, 1, 0}, \
        {0, 1, 0, 0, 0, 0, 1, 0}, \
        {0, 1, 0, 0, 0, 0, 1, 0}, \
        {0, 1, 0, 0, 0, 0, 1, 0}   \
}

#define E   { \
        {0, 1, 1, 1, 1, 1, 1, 0}, \
        {0, 1, 0, 0, 0, 0, 0, 0}, \
        {0, 1, 0, 0, 0, 0, 0, 0}, \
        {0, 1, 1, 1, 1, 1, 1, 0}, \
        {0, 1, 0, 0, 0, 0, 0, 0}, \
        {0, 1, 0, 0, 0, 0, 0, 0}, \
        {0, 1, 0, 0, 0, 0, 0, 0}, \
        {0, 1, 1, 1, 1, 1, 1, 0}   \
}

#define L { \
        {0, 1, 0, 0, 0, 0, 0, 0}, \
        {0, 1, 0, 0, 0, 0, 0, 0}, \
        {0, 1, 0, 0, 0, 0, 0, 0}, \
        {0, 1, 0, 0, 0, 0, 0, 0}, \
        {0, 1, 0, 0, 0, 0, 0, 0}, \
        {0, 1, 0, 0, 0, 0, 0, 0}, \
        {0, 1, 0, 0, 0, 0, 0, 0}, \
```

```
        {0, 1, 1, 1, 1, 1, 1, 0}   \
}

#define O { \
    {0, 0, 0, 1, 1, 0, 0, 0}, \
    {0, 0, 1, 0, 0, 1, 0, 0}, \
    {0, 1, 0, 0, 0, 0, 1, 0}, \
    {0, 1, 0, 0, 0, 0, 1, 0}, \
    {0, 1, 0, 0, 0, 0, 1, 0}, \
    {0, 1, 0, 0, 0, 0, 1, 0}, \
    {0, 0, 1, 0, 0, 1, 0, 0}, \
    {0, 0, 0, 1, 1, 0, 0, 0}   \
}

byte col = 0;
byte leds[8][8];

// pin[xx] on led matrix connected to nn on Arduino (-1 is dummy to make array start at pos 1)
int pins[17]= {-1,22,24, 26, 28, 30, 32,34,36,38,40,42,44,46,48,50,52};

// col[xx] of leds = pin yy on led matrix
int cols[8] = {pins[13], pins[3], pins[4], pins[10], pins[06], pins[11], pins[15], pins[16]};

// row[xx] of leds = pin yy on led matrix
int rows[8] = {pins[9], pins[14], pins[8], pins[12], pins[1], pins[7], pins[2], pins[5]};

const int numPatterns = 6;
byte patterns[numPatterns][8][8] = {
    H,E,L,L,O,SPACE
};

int pattern = 0;

void setup() {
    // sets the pins as output
    for (int i = 1; i <= 16; i++) {
        pinMode(pins[i], OUTPUT);
    }
```

```
// set up cols and rows
for (int i = 1; i <= 8; i++) {
   digitalWrite(cols[i - 1], turnoff);
}

for (int i = 1; i <= 8; i++) {
   digitalWrite(rows[i - 1], turnoff);
}

clearLeds();

// Turn off toggling of pin 11
FrequencyTimer2::disable();
// Set refresh rate (interrupt timeout period)
FrequencyTimer2::setPeriod(2000);
// Set interrupt routine to be called
FrequencyTimer2::setOnOverflow(display);

setPattern(pattern);
}

void loop() {
    pattern = ++pattern % numPatterns;
    slidePattern(pattern, 60);
}

void clearLeds() {
  // Clear display array
  for (int i = 0; i < 8; i++) {
    for (int j = 0; j < 8; j++) {
       leds[i][j] = 0;
    }
  }
}

void setPattern(int pattern) {
  for (int i = 0; i < 8; i++) {
    for (int j = 0; j < 8; j++) {
```

```
            leds[i][j] = patterns[pattern][i][j];
        }
    }
}

void slidePattern(int pattern, int del) {
    for (int l = 0; l < 8; l++) {
        for (int i = 0; i < 7; i++) {
            for (int j = 0; j < 8; j++) {
                leds[j][i] = leds[j][i+1];
            }
        }
        for (int j = 0; j < 8; j++) {
            leds[j][7] = patterns[pattern][j][0 + l];
        }
        delay(del);
    }
}

// Interrupt routine
void display() {
    digitalWrite(cols[col], HIGH);   // Turn whole previous column off
    col++;
    if (col == 8) {
        col = 0;
    }
    for (int row = 0; row < 8; row++) {
        if (leds[col][7 - row] == 1) {
            digitalWrite(rows[row], turnon);   // Turn on this led
        }
        else {
            digitalWrite(rows[row], turnoff); // Turn off this led
        }
    }
    digitalWrite(cols[col], turnoff); // Turn whole column on at once (for equal lighting
times)
}
```

讀者也可以在筆者 YouTube 頻道(https://www.youtube.com/user/UltimaBruce)中，在網址：https://www.youtube.com/watch?v=A0ABPJDS0tI&feature=youtu.be，看到本次實驗-滑動顯示 8x8 Led 點陣顯示器測試程式結果畫面，如下圖所示，可以看到 8x8 Led 點陣顯示器滑動顯示『HELLO』。

圖 68 滑動顯示 8x8 Led 點陣顯示器結果畫面

章節小結

本章主要介紹之 Arduino 開發板使用與使用單一個 Led 發光二極體，將之聯接為矩陣式 LED 顯示陣列，透過本章節的解說，相信讀者會對連接、控制大量 Led 發光二極體，有更深入的了解與體認。

CHAPTER

專屬驅動器之 Led 發光二極體顯示螢幕

本文實驗為了讓讀者可以更簡單驅動 Led，本文採用 MAX7219 來驅動點陣式 led，一般而言，我們直覺可以使用基本的 Led 發光二極體，由圖 58 所示，將之並列成排，由排成陣，組合成一個完整的 Led 陣列，或許讀者覺的圖 58 所示的 Led 沒有太複雜，由圖 58.(c)所示，共有 8*18 Led 排在一起，背後須要焊接 8*18 顆*2 針腳=>288 個焊接點，相信許多讀者看到就頭暈了，所以目前使用基本的 Led 發光二極體來排列成陣列的方式並不多見，因施工成本高、錯誤率高、維護成本高、易損壞、耗電量高等等眾多因素，所以越來越多人採用如下圖所示的矩陣式 LED(曹永忠, 許智誠, et al., 2014a, 2014b; 曹永忠, 許智誠, et al., 2014a, 2014b, 2014c, 2014d, 2014e, 2014f, 2015e, 2015k)。

如何控制矩陣式 LED

如果採用由圖 58.(c)所示，共有 8*18 Led 排在一起，背後須要焊接 8*18 顆*2 針腳=>288 個焊接點，要透過 Arduino 開發板來控制閃爍，需要 144 的 I/O 接點，不止如此，144 顆的 Led 發光二極體耗電量也是非常的驚人，更不是 Arduino 開發板可以支援的設計。

所以我們可以採用如下圖所示的矩陣式 LED，透過掃描式的方式來更新每一個 Led 點，透過人眼的視覺暫留⁸的特性與原理，所有的點在十分之一秒內可以全部輪流亮過，對人眼的視覺來說，自動會組合成一個完整的畫面，所以利用這樣的

---

⁸ 視覺暫留原理或稱視覺暫留作用、或視覺暫留性。視覺暫留是指人類的眼睛，對看到的物體或移動物體，約有 1/10 秒時間延遲與殘留，。如開燈 1/10 秒之後，我們才能看見物體；若關燈，1/10 秒之後，我們原看到的物體才會消失，因之，此種視覺延遲與殘留的現象，稱之視覺暫留。

原理來設計與控制矩陣式 LED，就可以拼湊出一個完整的畫面。

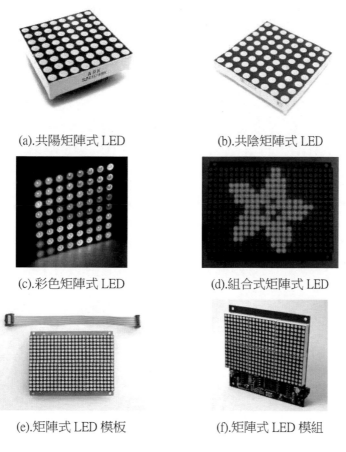

(a).共陽矩陣式 LED　　　　　　　(b).共陰矩陣式 LED

(c).彩色矩陣式 LED　　　　　　　(d).組合式矩陣式 LED

(e).矩陣式 LED 模板　　　　　　(f).矩陣式 LED 模組

圖 69 矩陣式 LED (Led Matrix)

　　但是矩陣式 LED 是製造廠商先行將 Led 置入如上圖所示的矩陣式 LED 顯示模組之中，並如下圖所示，將矩陣式 LED 所有的接腳整理成『列接腳』與『行接腳』。

　　如下圖(a).所示，只要先選定 R1~R7 之中任一腳，使其在低電位端，在透過程式將 C1~C5 的接腳，逐一設定要亮的點為高電位，第一階段的迴圈就是循序設定 R1~R7 的每一腳位為低電位，第二階段的迴圈就是設定 C1~C5 的每一腳位，逐一設定要亮的點為高電位，在兩層迴圈的程式架構下就可以完整的完成一個畫面的輸出(曹永忠, 許智誠, et al., 2014a; 曹永忠, 許智誠, et al., 2014a, 2014b, 2014c, 2014f)。

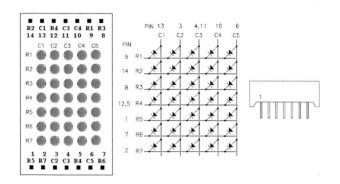

(a).7*5 矩陣式 LED 接腳圖

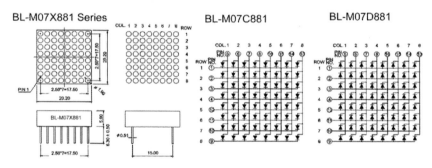

(b).8*8 矩陣式 LED 接腳圖

圖 70 矩陣式 LED 接腳圖

何謂 MAX7219

是 MAXIM 公司生產的整合的 IC，由下圖所示，MAX7219 是一個 24 接腳的 DIP 封裝或 DIP 包裝[9]，透過串列輸入/輸出共陰極顯示之驅動晶片(Driver IC)(詳細內

---

[9] 雙列直插封裝（dual in-line package） 也稱為 DIP 封裝或 DIP 包裝，簡稱為 DIP 或 DIL，是一種積體電路的封裝方式，積體電路的外形為長方形，在其兩側則有兩排平行的金屬接腳，稱為排針。DIP 包裝的元件可以焊接在印刷電路板電鍍的貫穿孔中，或是插入在 DIP 插座（socket）上。

容請參閱附錄資料),它可以連接微處理器與 8 位數字的 7 段數字 LED 顯示,也可以連接條線圖顯示器或者 64 個獨立的 LED:包括一個 B 型 BCD 編碼器、多路掃描迴路,還有一個 8*8 的靜態 RAM 用來存儲每一個數據。只需要一個外部暫存器用來設置各個 LED 的電流。

　　該晶片具有 10MHz 傳輸率的 SerialPeripheralinterface(SPI)三線串列10介面,並可以和任何微處理器(包括單晶片)相連,只需一個外接電阻即可設置所有 LED 的電流。它的操作很簡單,單晶片只需透過 SPI 三線介面就可以將相關的指令寫入 MAX7219 的內部指令和資料暫存器,同時它還允許使用者選擇多種解碼方式和譯碼位元。此外它還支援多片 7219 串聯方式,這樣單晶片就可以通過 3 根線(即串列資料線:Din、串列時鐘線:Clock 和晶片:CS 擇線)控制更多的 7 段顯示器或 8*8 矩陣式 LED 顯示器。

---

[10] Motorola 首先在其 MC68HCXX 系列處理器上定義 Serial Peripheral interface(SPI)介面主要應用在 EEPROM、FLASH、實時時鐘、AD 轉換器、還有數字信號處理器和數字信號解碼器之間。SPI,是一種高速的,全雙工,同步的通信,並且在晶片的接腳上只佔用四根線,節約了晶片的接腳,同時為 PCB 的佈局上節省空間,提供方便,正是出於這種簡單易用的特性,現在越來越多的晶片集成了這種通信協議。

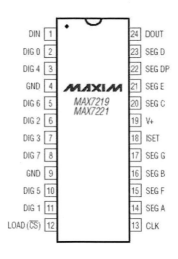

圖 71 Max7219 接腳一覽圖

Max7219 規格

為了進一步教導讀者使用 Max7219，我們將 Max7219 接腳說明如下：

- DIN：串列資料登錄端

- DOUT：串列資料輸出端，用於級連擴展

- LOAD：裝載資料登錄

- CLK：串列時鐘輸入

- DIG0~DIG7：8 位元 LED 位元選線，從共陰極 LED 中流入電流

- SEG A~SEG G DP 7 段顯式器驅動和小數點驅動

- ISET： 透過一個 10k 電阻和 Vcc 相連，設置段電流

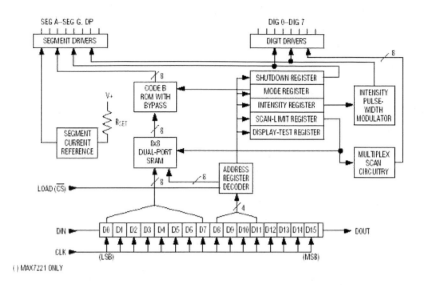

圖 72 MAX7219 的內部動作方塊圖

　　如下表所示，MAX7219 內部的暫存器，主要有：解碼控制暫存器、亮度控制暫存器、掃描界限暫存器、關斷模式暫存器、測試控制暫存器這幾個暫存器。

表 36 MAX7219 內部暫存器

Register Address Map

REGISTER	ADDRESS					HEX CODE
	D15–D12	D11	D10	D9	D8	
No-Op	X	0	0	0	0	0xX0
Digit 0	X	0	0	0	1	0xX1
Digit 1	X	0	0	1	0	0xX2
Digit 2	X	0	0	1	1	0xX3
Digit 3	X	0	1	0	0	0xX4
Digit 4	X	0	1	0	1	0xX5
Digit 5	X	0	1	1	0	0xX6
Digit 6	X	0	1	1	1	0xX7
Digit 7	X	1	0	0	0	0xX8
Decode Mode	X	1	0	0	1	0xX9
Intensity	X	1	0	1	0	0xXA
Scan Limit	X	1	0	1	1	0xXB
Shutdown	X	1	1	0	0	0xXC
Display Test	X	1	1	1	1	0xXF

我們介紹主要工作所需要的暫存器功能

解碼控制暫存器（X9H）

如下表所示，MAX7219 有兩種解碼方式：BCD 解碼方式和不解碼方式。當選擇不解碼時，8 個資料為分別一一對應 7 個段和小數點位；B 解碼方式是 BCD 解碼，直接送資料就可以顯示。

表 37 解碼控制暫存器（X9H）一覽表

DECODE MODE	REGISTER DATA								HEX CODE
	D7	D6	D5	D4	D3	D2	D1	D0	
No decode for digits 7–0	0	0	0	0	0	0	0	0	0x00
Code B decode for digit 0 No decode for digits 7–1	0	0	0	0	0	0	0	1	0x01
Code B decode for digits 3–0 No decode for digits 7–4	0	0	0	0	1	1	1	1	0x0F
Code B decode for digits 7–0	1	1	1	1	1	1	1	1	0xFF

掃描數暫存器（XBH）

如下表所示，此暫存器用於設置顯示的 LED 的個數（1~8），比如當設置為 0xX4 時，只會顯示 LED 0~5。

表 38 掃描數暫存器（XBH）一覽表

SCAN LIMIT	REGISTER DATA								HEX CODE
	D7	D6	D5	D4	D3	D2	D1	D0	
Display digit 0 only*	X	X	X	X	X	0	0	0	0xX0
Display digits 0 & 1*	X	X	X	X	X	0	0	1	0xX1
Display digits 0 1 2*	X	X	X	X	X	0	1	0	0xX2
Display digits 0 1 2 3	X	X	X	X	X	0	1	1	0xX3
Display digits 0 1 2 3 4	X	X	X	X	X	1	0	0	0xX4
Display digits 0 1 2 3 4 5	X	X	X	X	X	1	0	1	0xX5
Display digits 0 1 2 3 4 5 6	X	X	X	X	X	1	1	0	0xX6
Display digits 0 1 2 3 4 5 6 7	X	X	X	X	X	1	1	1	0xX7

亮度控制暫存器（XAH）

此暫存器控制顯示亮度大小，共有 16 個等級可選擇，用於設置 LED 的顯示亮度，暫存器的內容從 0xX0~0xXF。

斷模式暫存器（XCH）

此暫存器控制工作狀態的使用或關閉，共有兩種模式選擇，一是關閉狀態，（最低位元 D0=0）一是正常工作狀態（D0=1）。

顯示測試暫存器（XFH）

此暫存器控制 LED 控制狀態，使 Max7219 處在測試狀態還是正常工作狀態，當測試狀態時（最低位元 D0=1）各位顯示全亮，正常工作狀態(D0=0)。

讀寫時序說明

MAX7129 是 SPI 匯流排驅動方式。它不僅要向暫存器寫入控制字元，還需要讀取相應暫存器的資料。所以想要與 MAX7129 通訊，首先要先瞭解 MAX7129 的控制字元。MAX7129 的控制字元格式如圖 73 所示。

Serial-Data Format (16 Bits)

D15	D14	D13	D12	D11	D10	D9	D8	D7	D6	D5	D4	D3	D2	D1	D0
X	X	X	X	ADDRESS				MSB			DATA				LSB

圖 73 控制字元格式

工作時,MAX7219 規定一次接收 16 位元資料(Double Bytes)，在接收的 16 位元資料中：D15~D12 可以與操作無關，可以任意寫入，D11~D8 決定所選擇的內部暫存器的控制位址， D7~D0 為待顯示資料或是初始化控制字元。在 CLK 脈衝時序作用下，DIN 的資料以串列方式依次移入內部 16 位暫存器，然後在一個 LOAD 上升作用下，鎖存到內部的暫存器中。由下圖所示之 Max7219 工作時序圖注意在接收時，先接收最高位 D15，依序接收，最後是 D0，因此，在程式發送時必須先送高位元資料，在使迴圈方式來位移位元。

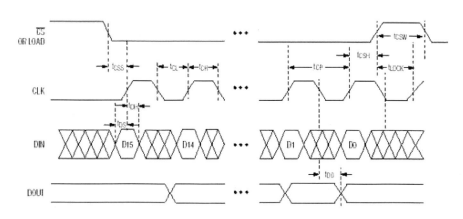

圖 74 Max7219 資料讀寫時序圖

電路原理圖

由下圖所示，Max7219 與單晶片的連接只需要 3 條線：LOAD（CS）晶片選擇接腳、CLK 串列時鐘接腳、DIN 串列資料接腳。

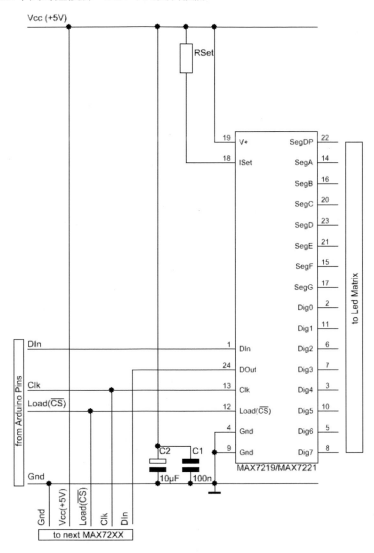

圖 75 MAX7219 驅動電路圖

資料來源：Arduino 官方網站：

http://playground.arduino.cc/Main/MAX72XXHardware#.Uzudz03NvmI

讀者對於 Max7219 連接 7 段顯示器或 8*8 矩陣式 LED 顯示器，可以參考 Arduino

官方網站，如下圖之 MAX7219 驅動 LED 電路圖，對於連接 7 段顯示器或 8*8 矩陣式 LED 顯示器都有電路範例可以參考。

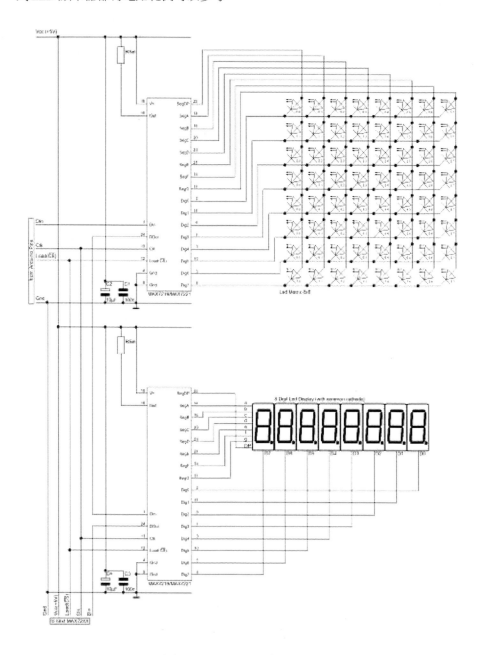

圖 76 MAX7219 驅動 LED 電路圖

資料來源：Arduino 官方網站：

http://playground.arduino.cc/uploads/Main/MAX72XX_Schematic.jpg

為了讓讀者更了解 Max7219 的運作原理，本文列舉下表之 8051 的 C 語言範例，

讓喜歡深入的讀者可以閱讀。

表 39 Max7219 之 8051 的 C 語言範例

Max7219 之 8051 的 C 語言範例(Max7219)

```c
sbit LOAD=P1^2;              //MAX7219 片選            12 腳
sbit DIN=P1^1;               //MAX7219 串列資料        1 腳
sbit CLK=P1^0;               //MAX7219 串列時鐘        13 腳
//暫存器巨集定義
#define DECODE_MODE    0x09      //解碼控制暫存器
#define INTENSITY      0x0A      //亮度控制暫存器
#define SCAN_LIMIT     0x0B      //掃描界限暫存器
#define SHUT_DOWN      0x0C      //關斷模式暫存器
#define DISPLAY_TEST   0x0F      //測試控制暫存器
//函數宣告
void Write7219(unsigned char address,unsigned char dat);
void Initial(void);
//位址、資料發送副程式
void Write7219(unsigned char address,unsigned char dat)
{
    unsigned char i;
    LOAD=0;        //拉低片選線，選中暫存器
    //發送地址
    for (i=0;i<8;i++)            //移位迴圈 8 次
    {
```

Max7219 之 8051 的 C 語言範例(Max7219)

```
    CLK=0;           //清零時鐘匯流排

    DIN=(bit)(address&0x80); //每次取高位元組

    address<<=1;                //左移一位

    CLK=1;           //時鐘上升沿，發送位址

}

//發送資料

for (i=0;i<8;i++)

{

    CLK=0;

    DIN=(bit)(dat&0x80);

    dat<<=1;

    CLK=1;           //時鐘上升沿，發送資料

}

LOAD=1;      //發送結束，上升沿鎖存資料

}

//MAX7219 初始化，設置 MAX7219 內部的控制暫存器

void Initial(void)

{

    Write7219(SHUT_DOWN,0x01);          //開啟正常工作模式（0xX1）

    Write7219(DISPLAY_TEST,0x00);       //選擇工作模式（0xX0）

    Write7219(DECODE_MODE,0xff);        //選用全解碼模式

    Write7219(SCAN_LIMIT,0x07);         //8 只 LED 全用

    Write7219(INTENSITY,0x04);          //設置初始亮度

}

測試程式
```

Max7219 之 8051 的 C 語言範例(Max7219)

```c
void main(void)
{
    unsigned char i;
    Initial();                //MAX7219 初始化
    while(1)
    {
        for(i=1;i<9;i++)
        {
            Write7219(i,i);         //數碼管顯示 1~8
        }
    }
}
```

Max7219 Led 顯示模組介紹

坊間有許多公司，如露天賣家：機械人 DIY 柑仔店

(http://class.ruten.com.tw/user/index00.php?s=cptc823)、柏毅電子

(http://class.ruten.com.tw/user/index00.php?s=boyi101)、安宸科技

(http://class.ruten.com.tw/user/index00.php?s=marktsai0316)或凱斯電子

(http://class.ruten.com.tw/user/index00.php?s=buyic)...等，都有販售 Max7219 相關模

組。

Max7219 IC 可以將 7 段顯示器或 8*8 矩陣式 LED 顯示器整合在一起，如

下圖所示，由於 Max7219 可以控制 7 段顯示器或 8*8 矩陣式 LED 顯示器，所

以可以見到下圖.(a)為 8*8 矩陣式 LED 顯示器模組，下圖.(b)為 7 段顯示器模組。

(a). 8*8 矩陣式 LED 顯示模組　　　　　　　　(b).7 段顯示器模組

圖 77 整合 Max7219 之顯示器模組

8*8 矩陣式 LED 顯示器模組產品規格

　　MAX7219 是一種整合的串列輸入/輸出共陰極顯示驅動器,它連接微處理器與 8 位元數位的 7 段數位 LED 顯示,也可以連接條線圖顯示器或者 64 個獨立的 LED。其上包括一個片上的 B 型 BCD 編碼器、多路掃描回路,段字驅動器,而且還有一個 8*8 的靜態 RAM 用來存儲每一個資料。只有一個外部暫存器用來設置各個 LED 的段電流。

　　只要使用四線串列介面可以聯接所有通用的微處理器。每個資料可以定址在更新時不需要改寫所有的顯示。MAX7219 同樣允許使用者對每一個資料選擇編碼或者不編碼。整個設備包含一個 150μA 的低功耗關閉模式,類比和數位亮度控制,一個掃描數量暫存器允許使用者顯示 1-8 位元資料,還有一個讓所有 LED 發光的檢測模式。

模組參數:

- 單個模組可以驅動一個 8*8 共陰 Led 矩陣顯示器
- 模組工作電壓:5V
- 模組尺寸: 5mm (長) X 3.2mm (寬) X 1.5mm(高)
- 帶 4 個固定螺絲孔,孔徑 3mm,可使用 M3 銅柱固定

● 模組帶輸入輸出介面，支援多個模組級聯

接線說明：

● 模組左邊為輸入埠，右邊為輸出埠。

● 控制單個模組時，只需要將輸入埠接到微處理機(單晶片)

● 多個模組級聯時，第 1 個模組的輸入端接微處理機(單晶片)，輸出端接第 2 個模組的輸入端，第 2 個模組的輸出端接第 3 個模組的輸入端，以此類推...

輸入端接腳說明：

● VCC → 5V

● GND → GND

● DIN → Data In Pin

● CS → Chip Select Pin COntrol

● CLK → Clock Pulse Control

使用 Max7219 Led 顯示模組

本章使用的 Max7219 Led 顯示模組(詳細資料可參閱附錄)，由上圖(a)所示，其連接電路非常簡單，若讀者想要其它連接方法的電路圖，可以參考『Arduino 實作布手環:Using Arduino to Implementation a Mr. Bu Bracelet』(曹永忠, 許智誠, et al., 2015k)、『Arduino 实作布手环:Using Arduino to Implementation a Mr. Bu Bracelet』(曹永忠, 許智誠, et al., 2015e)等書附錄章節中：Max7219 Led 模組線路圖』、『MAX7219 資料』等相關資料(曹永忠, 2016h, 2016i)。

本章節使用的 LedControl 函式庫，乃是 Eberhard Fahle 在其 github 網站分享函

式庫，讀者可以到 https://github.com/wayoda/LedControl 下載其函式庫，特感謝 Eberhard Fahle 提供。

讀者也可以到筆者 Github(https://github.com/)網站，本書的所有範例檔，都可以在 https://github.com/brucetsao/eLed，下載所需要的檔案。

首先，請讀者依照下表進行 Max7219 Led 模組電路組立，再進行程式攥寫的動作。

表 40 Max7219 Led 模組接腳表

	模組接腳	Arduino 開發板接腳	解說
Max7219 Led 模組	Vcc	Arduino +5V	Max7219 Led 模組 資料接腳
	GND	Arduino GND(共地接點)	
	DIN	Arduino Pin 43	
	CS	Arduino Pin 45	
	CLK	Arduino Pin 47	

我們，請讀者鍵入Ｓｋｅｔｃｈ　ＩＤＥ軟體(軟體下載請到：https://www.arduino.cc/en/Main/Software)，編譯完成後上傳到開發版進行測試。

完成 Arduino 開發板與 Max7219 Led 模組連接之後，將下列表 41 之 Max7219 矩陣式 LED 模組測試程式一鍵入 Arduino Sketch 之中，請讀者鍵入Ｓｋｅｔｃｈ　ＩＤＥ軟體(軟體下載請到：https://www.arduino.cc/en/Main/Software)，編譯完成後上傳到開發版進行測試。

如下圖所示，可以見到顯示幕上顯示英文字『A』。

表 41 Max7219 矩陣式 LED 模組測試程式一

Max7219 矩陣式 LED 模組測試程式一(max7219_led01)

```cpp
#include "LedControl.h"

/*

  Now we need a LedControl to work with.

  ***** These pin numbers will probably not work with your hardware *****

  pin 12 is connected to the DataIn

  pin 11 is connected to the CLK

  pin 10 is connected to LOAD

  We have only a single MAX72XX.

*/

#define DinPin 43

#define CsPin 45

#define ClkPin 47

#define chipno 0

LedControl lc=LedControl(DinPin,ClkPin,CsPin,chipno);


/* we always wait a bit between updates of the display */

unsigned long delaytime=2000;


void setup() {
  /*

    The MAX72XX is in power-saving mode on startup,

    we have to do a wakeup call

  */

  lc.shutdown(0,false);

  /* Set the brightness to a medium values */
```

```
  lc.setIntensity(0,8);

  /* and clear the display */

  lc.clearDisplay(0);

  Serial.begin(9600);

  Serial.println("Program Start Here");

}


void loop() {

  writeArduinoOnMatrix();

  //rows();

 // columns();

//   single();

}


/*

  This method will display the characters for the

  word "Arduino" one after the other on the matrix.

  (you need at least 5x7 leds to see the whole chars)

  */

void writeArduinoOnMatrix() {

  /* here is the data for the characters */

  byte a[8]={B00011000,

             B00100100,

             B01000010,

             B10000001,
```

```
            B11111111,

            B10000001,

            B10000001,

            B00000000};




    /* now display them one by one with a small delay */
    showhar(&a[0]);
      Serial.println("Now print (A)");
    delay(delaytime);



}
void showhar(byte *chr)
{
 int i = 0 ;
    clearchar();
    //lc.clearDisplay(0) ;
 for(i = 0 ; i <8 ; i++)
 {
    lc.setColumn(0,i,*(chr+i));
 }


}
```

```
void clearchar()

{

 int i = 0 ;

 for(i = 0 ; i <8 ; i++)

 {

    lc.setColumn(0,i,0x00);

 }


}


/*

   This fushoion lights ua some Leds in a row.

   The pattern will be repeatd on every row.

   The pattern will blink along with the row-number.

   row number 4 (index==3) will blink 4 times etc.

   */

void rows() {

   for(int row=0;row<8;row++) {

      delay(delaytime);

      lc.setRow(0,row,B10100000);

      delay(delaytime);

      lc.setRow(0,row,(byte)0);

      for(int i=0;i<row;i++) {

         delay(delaytime);
```

```
        lc.setRow(0,row,B10100000);

        delay(delaytime);

        lc.setRow(0,row,(byte)0);

    }

  }

}


/*

  This function lights up a some Leds in a column.

  The pattern will be repeated on every column.

  The pattern will blink along with the column-number.

  column number 4 (index==3) will blink 4 times etc.
*/
void columns() {
  for(int col=0;col<8;col++) {

    delay(delaytime);

    lc.setColumn(0,col,B10100000);

    delay(delaytime);

    lc.setColumn(0,col,(byte)0);

    for(int i=0;i<col;i++) {

      delay(delaytime);

      lc.setColumn(0,col,B10100000);

      delay(delaytime);

      lc.setColumn(0,col,(byte)0);

    }
```

```
    }

}


/*

This function will light up every Led on the matrix.

The led will blink along with the row-number.

row number 4 (index==3) will blink 4 times etc.

*/

void single() {

   for(int row=0;row<8;row++) {

      for(int col=0;col<8;col++) {

         delay(delaytime);

         lc.setLed(0,row,col,true);

         delay(delaytime);

         for(int i=0;i<col;i++) {

            lc.setLed(0,row,col,false);

            delay(delaytime);

            lc.setLed(0,row,col,true);

            delay(delaytime);

         }

      }

   }

}
```

圖 78 Max7219 矩陣式 LED 模組測試程式一結果畫面

進階使用 Max7219 Led 顯示模組

若我們每輸出一個字元，就必需重新定義一個字元陣列，這樣使用上非常不方便，所以我們可以一開始就將用到的字元陣列讀入，使用 ASCII 的方式來驅動顯示字元，相信會更加容易(曹永忠, 2016h, 2016i)。

完成 Arduino 開發板與 Max7219 Led 模組連接之後，將下表所示之 Max7219 矩陣式 LED 模組測試程式二，請讀者鍵入Ｓｋｅｔｃｈ　ＩＤＥ軟體(軟體下載請到：https://www.arduino.cc/en/Main/Software)，編譯完成後上傳到開發版進行測試。

如下圖所是，可以見到把 96 個預先定義的字元一一顯示出來。

表 42 Max7219 矩陣式 LED 模組測試程式二

Max7219 矩陣式 LED 模組測試程式二(max7219_led02)
#include <LedControl.h> /* Now we need a LedControl to work with. ***** These pin numbers will probably not work with your hardware *****

- 217 -

```
pin 12 is connected to the DataIn

pin 11 is connected to the CLK

pin 10 is connected to LOAD

We have only a single MAX72XX.

*/

#define DinPin 43

#define CsPin 45

#define ClkPin 47

#define chipno 0

LedControl lc=LedControl(DinPin,ClkPin,CsPin,chipno);


/* we always wait a bit between updates of the display */

unsigned long delaytime=2000;


byte alphabets[][5] = {

    {0x00,0x00,0x00,0x00,0x00},  /*space*/ // is 32 in ASCII

    {0x00,0xF6,0xF6,0x00,0x00},  /*!*/

    {0x00,0xE0,0x00,0xE0,0x00},  /*"*/

    {0x28,0xFE,0x28,0xFE,0x28},  /*#*/

    {0x00,0x64,0xD6,0x54,0x08},  /*$*/

    {0xC2,0xCC,0x10,0x26,0xC6},  /*%*/

    {0x4C,0xB2,0x92,0x6C,0x0A},  /*&*/

    {0x00,0x00,0xE0,0x00,0x00},  /*'*/

    {0x00,0x38,0x44,0x82,0x00},  /*(*/

    {0x00,0x82,0x44,0x38,0x00},  /*)*/
```

```
{0x88,0x50,0xF8,0x50,0x88},    /***/

{0x08,0x08,0x3E,0x08,0x08},    /*+*/

{0x00,0x00,0x05,0x06,0x00},    /*,*/

{0x08,0x08,0x08,0x08,0x08},    /*-*/

{0x00,0x00,0x06,0x06,0x00},    /*.*/

{0x02,0x0C,0x10,0x60,0x80},    /*/*/

{0x7C,0x8A,0x92,0xA2,0x7C},    /*0*/

{0x00,0x42,0xFE,0x02,0x00},    /*1*/

{0x42,0x86,0x8A,0x92,0x62},    /*2*/

{0x44,0x82,0x92,0x92,0x6C},    /*3*/

{0x10,0x30,0x50,0xFE,0x10},    /*4*/

{0xE4,0xA2,0xA2,0xA2,0x9C},    /*5*/

{0x3C,0x52,0x92,0x92,0x0C},    /*6*/

{0x80,0x86,0x98,0xE0,0x80},    /*7*/

{0x6C,0x92,0x92,0x92,0x6C},    /*8*/

{0x60,0x92,0x92,0x94,0x78},    /*9*/

{0x00,0x00,0x36,0x36,0x00},    /*:*/

{0x00,0x00,0x35,0x36,0x00},    /*;*/

{0x10,0x28,0x44,0x82,0x00},    /*<*/

{0x28,0x28,0x28,0x28,0x28},    /*=*/

{0x00,0x82,0x44,0x28,0x10},    /*>*/

{0x40,0x80,0x8A,0x90,0x60},    /*?*/

{0x7C,0x82,0xBA,0xBA,0x62},    /*@*/

{0x3E,0x48,0x88,0x48,0x3E},    /*A*/

{0xFE,0x92,0x92,0x92,0x6C},    /*B*/
```

```
{0x7C,0x82,0x82,0x82,0x44},   /*C*/

{0xFE,0x82,0x82,0x82,0x7C},   /*D*/

{0xFE,0x92,0x92,0x92,0x82},   /*E*/

{0xFE,0x90,0x90,0x90,0x80},   /*F*/

{0x7C,0x82,0x82,0x8A,0x4E},   /*G*/

{0xFE,0x10,0x10,0x10,0xFE},   /*H*/

{0x82,0x82,0xFE,0x82,0x82},   /*I*/

{0x84,0x82,0xFC,0x80,0x80},   /*J*/

{0xFE,0x10,0x28,0x44,0x82},   /*K*/

{0xFE,0x02,0x02,0x02,0x02},   /*L*/

{0xFE,0x40,0x20,0x40,0xFE},   /*M*/

{0xFE,0x60,0x10,0x0C,0xFE},   /*N*/

{0x7C,0x82,0x82,0x82,0x7C},   /*O*/

{0xFE,0x90,0x90,0x90,0x60},   /*P*/

{0x7C,0x82,0x82,0x86,0x7E},   /*Q*/

{0xFE,0x90,0x98,0x94,0x62},   /*R*/

{0x64,0x92,0x92,0x92,0x4C},   /*S*/

{0x80,0x80,0xFE,0x80,0x80},   /*T*/

{0xFC,0x02,0x02,0x02,0xFC},   /*U*/

{0xF8,0x04,0x02,0x04,0xF8},   /*V*/

{0xFC,0x02,0x0C,0x02,0xFC},   /*W*/

{0xC6,0x28,0x10,0x28,0xC6},   /*X*/

{0xC0,0x20,0x1E,0x20,0xC0},   /*Y*/

{0x86,0x8A,0x92,0xA2,0xC2},   /*Z*/

{0x00,0x00,0xFE,0x82,0x00},   /*[*/
```

```
{0x00,0x00,0x00,0x00,0x00},    /*this should be / */

{0x80,0x60,0x10,0x0C,0x02},    /*]*/

{0x20,0x40,0x80,0x40,0x20},    /*^*/

{0x02,0x02,0x02,0x02,0x02},    /*_*/ // use for 7 dot high display

{0x80,0x40,0x20,0x00,0x00},    /*`*/

{0x04,0x2A,0x2A,0x2A,0x1E},    /*a*/

{0xFE,0x12,0x22,0x22,0x1C},    /*b*/

{0x1C,0x22,0x22,0x22,0x14},    /*c*/

{0x1C,0x22,0x22,0x12,0xFE},    /*d*/

{0x1C,0x2A,0x2A,0x2A,0x18},    /*e*/

{0x10,0x7E,0x90,0x80,0x40},    /*f*/

{0x18,0x25,0x25,0x25,0x1E},    /*g*/

{0xFE,0x10,0x10,0x10,0x0E},    /*h*/

{0x00,0x12,0x5E,0x02,0x00},    /*i*/

{0x02,0x01,0x01,0x11,0x5E},    /*j*/

{0xFE,0x08,0x08,0x14,0x22},    /*k*/

{0x00,0x82,0xFE,0x02,0x00},    /*l*/

{0x3E,0x20,0x1C,0x20,0x1E},    /*m*/

{0x3E,0x20,0x20,0x20,0x1E},    /*n*/

{0x1C,0x22,0x22,0x22,0x1C},    /*o*/

{0x3F,0x24,0x24,0x24,0x18},    /*p*/

{0x18,0x24,0x24,0x3F,0x01},    /*q*/

{0x3E,0x10,0x20,0x20,0x10},    /*r*/

{0x12,0x2A,0x2A,0x2A,0x04},    /*s*/

{0x00,0x10,0x3C,0x12,0x04},    /*t*/
```

```
  {0x3C,0x02,0x02,0x02,0x3E},    /*u*/

  {0x30,0x0C,0x02,0x0C,0x30},    /*v*/

  {0x38,0x06,0x18,0x06,0x38},    /*w*/

  {0x22,0x14,0x08,0x14,0x22},    /*x*/

  {0x38,0x05,0x05,0x05,0x3E},    /*y*/

  {0x22,0x26,0x2A,0x32,0x22},    /*z*/

  {0x00,0x10,0x6C,0x82,0x82},    /*{*/

  {0x00,0x00,0xFF,0x00,0x00},    /*|*/

  //{0x04,0x02,0xFF,0x02,0x04},    /*|, down arrow*/

  {0x82,0x82,0x6C,0x10,0x00},    /*}*/

  {0x08,0x10,0x18,0x08,0x10},    /*~*/

};


void setup() {
  /*

  The MAX72XX is in power-saving mode on startup,

  we have to do a wakeup call

  */
  lc.shutdown(0,false);
  /* Set the brightness to a medium values */
  lc.setIntensity(0,8);
  /* and clear the display */
  lc.clearDisplay(0);
  Serial.begin(9600);
  Serial.println("Program Start Here");
```

```
}


void loop() {
 listallalpha(2000);

}



void showalpha(char w)

{
   byte aa;


   aa = w - 32;
   Serial.print("now print (");
   Serial.print(aa,DEC);
   Serial.println(")");


   showchar(&(alphabets[aa][0]));
}
void showchar(byte *chr)

{
 int i = 0 ;
   //clearchar();
   lc.clearDisplay(chipno);
 for(i = 0 ; i <5 ; i++)

 {
```

```
    lc.setColumn(chipno,i,*(chr+i));

    Serial.print(*(chr+i),HEX);

}

    Serial.println("/");


}

void listallalpha(int delaytime)

{

    byte ii = 0 ;

  for(ii = 0 ; ii <=96; ii++)

  {

    showchar(&(alphabets[ii][0]));

    delay(1000);

  }


  delay(delaytime);

}

void clearchar()

{

 int i = 0 ;

 for(i = 0 ; i <8 ; i++)

 {

    lc.setColumn(chipno,i,0x00);

 }
```

Max7219 矩陣式 LED 模組測試程式二(max7219_led02)
}

整合多字幕型 Max7219 Led 顯示模組

由於單一 Max7219 Led 顯示模組只能顯示 8*8 個點，約列只能顯示一個英文字元，實在稱不上一個字幕機，但是 Max7219 Led 顯示模組可以透過串接的方式，目前 Max7219 Led 顯示模組可以串接到 8 個 Max7219 Led 顯示模組，筆者向網路露天賣家安宸科技(http://class.ruten.com.tw/user/index00.php?s=marktsai0316)，購買共 四 個 Max7219 Led 顯 示 模 組 (http://goods.ruten.com.tw/item/show?21304113897100) ，賣家安宸科技也提供修改版的 LedCOntrol 函式庫，讀者可以向該賣家購買模組時，記得向賣家安宸科技索取修改版的 LedCOntrol 函式，該修改版的 LedCOntrol 函式庫可參考本書附錄『LedControl 函式庫(安宸科技版)』，特此感謝賣家安宸科技的分享。

我們由下圖所示，由於 Max7219 IC 本身的設計與架構，本身就可以串聯，所以有廠商就製造串聯型 Max7219 Led 顯示模組，由於裝上所有零件的串聯型 Max7219 Led 顯示模組，不易看見輸入與輸出的接腳，所以我們透過下圖.(a).所示，在 Max7219 IC 左邊為訊號輸入端，在點陣式 Led 顯示器(參考下圖.(b). Max7219 模組零件包)右方，有同樣的訊號排列的訊號輸出端，所以第二組的 Max7219 Led 顯示模組其訊號輸入端接在第一組的 Max7219 Led 顯示模組其訊號輸出端，第三組的 Max7219 Led 顯示模組其訊號輸入端接在第二組的 Max7219 Led 顯示模組其訊號輸出端，以此類推。

(a).未插上零件之 Max7219 Led 顯示模組　　　　(b). Max7219 模組零件包

圖 79 串聯型 Max7219 Led 顯示模組

由上述，我們可以參考下表所示，將四組的 Max7219 Led 顯示模組串接起來(曹永忠, 2016h, 2016i)。

表 43 多字幕型 Max7219 Led 模組接腳表

	模組接腳	Arduino 開發板接腳	解說
Max7219 Led (a)	Vcc	Arduino +5V	Max7219 Led 模組 第一組 資料接腳
	GND	Arduino GND(共地接點)	
	DIN	Arduino Pin 43	
	CS	Arduino Pin 45	
	CLK	Arduino Pin 47	
Max7219 Led (b)	Vcc	第一組輸出 Vcc	Max7219 Led 模組 第二組 資料接腳
	GND	第一組輸出 GND	
	DIN	第一組輸出 DIN	
	CS	第一組輸出 CS	
	CLK	第一組輸出 CLK	
Max7219 Led (c)	Vcc	第二組輸出 Vcc	Max7219 Led 模組 第三組 資料接腳
	GND	第二組輸出 GND	
	DIN	第二組輸出 DIN	
	CS	第二組輸出 CS	
	CLK	第二組輸出 CLK	
Max	Vcc	第三組輸出 Vcc	Max7219 Led 模組
	GND	第三組輸出 GND	

DIN	第三組輸出 DIN	第四組
CS	第三組輸出 CS	
CLK	第三組輸出 CLK	資料接腳

完成 Arduino 開發板與四組 Max7219 Led 顯示模組串接之後，將下表所示之串聯型 Max7219 Led 顯示模組測試程式一，請讀者鍵入Ｓｋｅｔｃｈ　ＩＤＥ軟體 (軟體下載請到：https://www.arduino.cc/en/Main/Software)，編譯完成後上傳到開發版進行測試。

如下圖所示，我們可以見到英文字串『Hello World!』可以在四組的 Max7219 Led 顯示模組依序顯示，並可以平滑移動與捲軸。

表 44 串聯型 Max7219 Led 顯示模組測試程式一

串聯型 Max7219 Led 顯示模組測試程式一(max7219_led03)

```
#include <LedControl.h>


//We always have to include the library
//#include "LedControl.h"


/*

 Now we need a LedControl to work with.

 ***** These pin numbers will probably not work with your hardware *****

 pin 12 is connected to the DataIn

 pin 10 is connected to LOAD

 pin 11 is connected to the CLK

 We have only a single MAX72XX.

 */
```

```
#define DinPin 43

#define CsPin 45

#define ClkPin 47

#define chipno 0

#define chips 8


//LedControl lc=LedControl(12,10,11,9);

LedControl lc=LedControl(DinPin,ClkPin,CsPin,chips);

/* we always wait a bit between updates of the display */

unsigned long delaytime=1000;


void setup() {
  /*
    The MAX72XX is in power-saving mode on startup,
    we have to do a wakeup call
  */
  for(int i=0;i<4;i++)
  {
  lc.shutdown(i,false);
  /* Set the brightness to a medium values */
  lc.setIntensity(i,8);
  /* and clear the display */
  lc.clearDisplay(i);
  }
```

```
  writeArduinoOnMatrix();

}


/*

This method will display the characters for the

word "Arduino" one after the other on the matrix.

(you need at least 5x7 leds to see the whole chars)

*/

void writeArduinoOnMatrix() {

  /* here is the data for the characters */

  byte a[8]={B00000000,

             B00000000,

             B01000010,

             B01111110,

             B01000010,

             B00000000,

             B00000000,

             B00000000};

  byte b[8]={B00000000,

             B00011100,

             B00100010,

             B01000100,

             B01000100,

             B00100010,

             B00011100,
```

```
                              B00000000};

byte c[8]={B00000000,

                              B00000010,

                              B00000010,

                              B01111110,

                              B00000010,

                              B00000010,

                              B00000000,

                              B00000000};

byte d[8]={B00000000,

                              B00111100,

                              B01000010,

                              B01000010,

                              B01000010,

                              B00100100,

                              B00000000,

                              B00000000};

byte e[8]={B00000000,

                              B00000010,

                              B00000010,

                              B01111110,

                              B00000010,

                              B00000010,

                              B00000000,

                              B00000000};
```

```
byte f[8]={B00000000,
            B01111110,
            B00000100,
            B00011000,
            B00000100,
            B01111110,
            B00000000,
            B00000000};


/* now display them one by one with a small delay */
//    lc.setCharFont5X7(0, 0, 'H');
//    lc.setCharFont5X7(1, 0, 'e');
    lc.setStringFont5X7(0,0,"Hello World!",6); //FontWidth=6
#if 0
    lc.setRow(0,0,a[0]);
    lc.setRow(0,1,a[1]);
    lc.setRow(0,2,a[2]);
    lc.setRow(0,3,a[3]);
    lc.setRow(0,4,a[4]);
    lc.setRow(0,5,a[5]);
    lc.setRow(0,6,a[6]);
    lc.setRow(0,7,a[7]);
    lc.setRow(1,0,b[0]);
    lc.setRow(1,1,b[1]);
    lc.setRow(1,2,b[2]);
```

```
    lc.setRow(1,3,b[3]);

    lc.setRow(1,4,b[4]);

    lc.setRow(1,5,b[5]);

    lc.setRow(1,6,b[6]);

    lc.setRow(1,7,b[7]);
#endif
    Serial.begin(9600);

    Serial.println(lc.status[0],BIN);

    Serial.println(lc.status[1],BIN);

    Serial.println(lc.status[2],BIN);

    Serial.println(lc.status[3],BIN);

    Serial.println(lc.status[4],BIN);

    Serial.println(lc.status[5],BIN);

    Serial.println(lc.status[6],BIN);

    Serial.println(lc.status[7],BIN);


//   char *a=new char(5);

}


void LShift()

{

    byte firststatus=lc.status[0];

    lc.setRow(0,0,lc.status[1]);

    lc.setRow(0,1,lc.status[2]);

    lc.setRow(0,2,lc.status[3]);
```

```
lc.setRow(0,3,lc.status[4]);

lc.setRow(0,4,lc.status[5]);

lc.setRow(0,5,lc.status[6]);

lc.setRow(0,6,lc.status[7]);

 lc.setRow(0,7,lc.status[8]);

 lc.setRow(1,0,lc.status[9]);

lc.setRow(1,1,lc.status[10]);

lc.setRow(1,2,lc.status[11]);

lc.setRow(1,3,lc.status[12]);

lc.setRow(1,4,lc.status[13]);

lc.setRow(1,5,lc.status[14]);

lc.setRow(1,6,lc.status[15]);

lc.setRow(1,7,firststatus);

delay(100);


}


void loop() {

//LShift();//   writeArduinoOnMatrix();

lc.LeftRotate(1);

//delay(100);

}
```

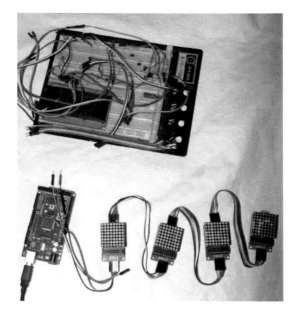

圖 80 串聯型 Max7219 Led 顯示模組測試程式一結果畫面

LedControl 函數用法

為了更能了解 LedControl 的用法，本節詳細介紹了 LedControl 函式主要的用法：

物件產生

LedControl mydisplay = LedControl(DIN_pin, CLK_pin, LOAD_pin, number_of_chips);

1. 指令格式 LedControl(DIN_pin, CLK_pin, LOAD_pin, number_of_chips);
2. 使用參數個格式如下：
   - DIN_pin：Din 腳位設定
   - CLK_pin：Clock 時脈控制腳位設定
   - LOAD_pin：Chip Select:CS 晶片選擇腳位設定

- number_of_chips：共有幾片 Max7219 Led 顯示模組串接(1~8 組)

物件設定(Setup)

LedControl.shutdown(chip, false);

啟動或關閉 Max7219 Led 顯示模組

1. 使用參數個格式如下：

   甲、chip：第幾個模組

   乙、false：啟動 Max7219

   丙、True：關閉 Max7219

2. 指令範例：

   LedControl.shutdown(0, false);

   　　　啟動第一組 Max7219 Led 顯示模組

LedControl. setIntensity(chip, intensity);

設定 Max7219 Led 顯示亮度強度

1. 使用參數個格式如下：

   甲、chip：第幾個模組

   乙、intensity：0~15，設定顯示亮度強度

2. 指令範例：

   setIntensity(0, 15);;

   　　　設定第一組 Max7219 Led 顯示亮度為 15(最亮)

LedControl. getDeviceCount();

回傳共有幾個 Max7219 Led 顯示模組

LedControl. setScanLimit(int addr, int limit);

設定 Max7219 Led 控制多少燈

1. 使用參數個格式如下：

甲、addr：第幾個模組

乙、limit：0~7，設定控制多少燈

七段顯示器顯示控制(Display Function)

LedControl. clearDisplay(int addr)

清除 Max7219 Led 顯示模組內容

1. 使用參數個格式如下：

甲、addr：第幾個模組

LedControl.setDigit(chip, digit_position, number, dot);;

設定 Max7219 Led 顯示模組使用七段顯示器數字時某一個七段顯示器數字內容

1. 使用參數個格式如下：

甲、chip：第幾個模組

乙、digit_position：第 digit_position 個七段顯示器

丙、number ：第 digit_position 個七段顯示器的內容值(Byte 型態)

丁、dot：內容為 true；顯示小數點，內容為 false；不顯示小數點

2. 指令範例：

LedControl.setDigit(0,2,(byte)3,false);

LedControl.setDigit(0,1,(byte)2,true);

LedControl.setDigit(0,0,(byte)4, false);

顯示 32.4

LedControl. setLed(chip, row, column, state);

設定 Max7219 Led 顯示模組使用七段顯示器數字時某一個七段顯示器那一段
燈管亮或不亮

1. 使用參數個格式如下：

甲、chip：第幾個模組

乙、row：第 row 個七段顯示器

丙、column：第 column 個段燈管

丁、state：內容為 true；燈管亮，內容為 false；燈管不亮

2. 指令範例：

LedControl.setLed(0,2,7,true);

設定第一組 Max7219 Led 顯示模組的七段顯示器第二個數字第七段
燈管亮

LedControl. setRow(int addr, int row, byte value);

設定 Max7219 Led 顯示模組使用 Led 陣列顯示模組第 N 列的內容

1. 使用參數個格式如下：

甲、addr：第幾個模組

乙、row：第 row 列

丙、value：第 row 列陣列顯示模組的內容值(Byte 型態)

2. 指令範例：
LedControl. setRow(0,2,146);
顯示第 2 列 10010010(1 表亮，0 表不亮)

LedControl. setColumn(int addr, int col, byte value);

設定 Max7219 Led 顯示模組使用 Led 陣列顯示模組第 N 行的內容

1. 使用參數個格式如下：

甲、addr：第幾個模組

乙、col：第 col 行

丙、value：第 col 行陣列顯示模組的內容值(Byte 型態)

2. 指令範例：

LedControl. setColumn (0,2,146);
顯示第 2 行 0010010(1 表亮，0 表不亮)

LedControl.setChar(int addr, int digit, char value, boolean dp);

設定 Max7219 Led 顯示模組使用七段顯示器數字時某一個七段顯示器數字內

容

1. 使用參數個格式如下：

甲、addr：第幾個模組

乙、digit：第 digit 個七段顯示器

丙、value：第 digit 個七段顯示器的內容值(Byte 型態)

丁、dp：內容為 true；顯示小數點，內容為 false；不顯示小數點

2. 指令範例：

setChar(0,3,'-',false););

顯示『 - 』

章節小結

本章主要介紹之 Arduino 開發板使用與連接 Max7219 矩陣式 LED 模組，透過

本章節的解說，相信讀者會對連接、使用 Max7219 矩陣式 LED 模組，有更深入的
了解與體認。

8

CHAPTER

彩色矩陣式 LED 顯示模組

本文介紹彩色矩陣式 LED 顯示模組，市面上可以購得此類商品化的產品，本實驗彩色字幕機購自露天拍賣賣家微控科技 (http://class.ruten.com.tw/user/index00.php?s=kiwiapple77)的 Colorduino Shield (商家產品名稱為:RGB 控制底板 + 彩色 8x8 點陣 LED / Arduino IDE 直接燒寫 1600 萬色產品)(拍賣網址：http://goods.ruten.com.tw/item/show?21302192412175)，如下圖所示，為完整的 Colorduino Shield 全圖: 下圖.(a).為 RGB Led 8*8 點陣模組；下圖.(b).為 Colorduino Shield 電路板一覽圖；下圖.(c).為 Colorduino Shield 正視圖；下圖(d).為 Colorduino Shield 背視圖(曹永忠、許智誠, et al., 2014a; 曹永忠, 許智誠, et al., 2014a)。

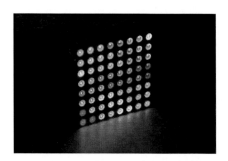

(a). RGB 8*8 點陣模組

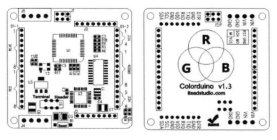

(b). Colorduino Shield 電路板一覽圖

(c). Colorduino Shield 正視圖   (d). Colorduino Shield 背視圖

圖 81 Colorduino Shield 模組

Colorduino Shield

Colorduino 是由 ITead Studio[11]在 Arduino 架構基礎上，專門用於驅動全彩 RGB LED 點陣的驅動板。Colorduino 使用 DM163[12]作為 RGB 8*8 點陣模組驅動晶片，最多支持全彩 Red Grey:256xGreen Grey:256xBlue Grey:256　= 16,777,216 色。

Colorduino 支持多塊 Colorduino Shield 模組串聯，然後通過 IIC 通訊方式來控制 Colorduino Shield 模組。

Colorduino Shield 模組產品特點

- 顏色支援全彩 16,777,216 色

- 支持 16MHZ PWM 相關硬體

- 可不需任何外加的電路，可獨立運作

- 支持 URAT 串列通訊和 IIC 通信方式

- 通道電流可達 100mA

- 驅動電流達 500mA

---

[11] http://blog.iteadstudio.com/

[12] http://www.siti.com.tw/product/spec/LED/DM163.pdf

該 Colorduino Shield 模組可搭配圖 81.(a).為 RGB Led 8*8 點陣模組，其 RGB Led 8*8 點陣模組規格如下表所示：

表 45 RGB 8*8 點陣模組

規格	內容
Dot Size	5.0mm
Pixel Array	8×8
Luminous Intensity	40mcd
Package Dimension	60mm×60mm
Reverse Voltage(Max)	5V
Forward Current(Max)	25mA
Peak Forward Current(Max)	100mA
Power Dissipation(Max)	100mW
Operating Temperature(Max)	-35~+85℃
Storage Temperature(Max)	-35~+85℃
Lead Solder Temperature(Max)	-260℃ for 5 seconds
Emitted color	RGB
LED Color	Red-Green-Blue
Matrix Type	Common Anode
Dot Size	5mm
Pixel Array	8×8
Luminous Intensity	40mced
Package Dimension	60mm×60mm
Reverse Voltage	5V
Forward Current	25mA
Weight	45.00g
Model	IM120601007

驅動 Colorduino

Colorduino Shield 模組本身就有內含 Atmel 微控制器 ATmega328，透過 DM163 全彩控制晶片，可以自主的驅動 8*8 的 RGB Led(曹永忠, 許智誠, et al., 2014a; 曹永忠, 許智誠, et al., 2014a)，所以本身可以不需 Arduino 開發板就可以自主驅動。

所以我們採用 USB2TTL，連接 Colorduino Shield 模組，讀者依照下表所示之 Colorduino 接腳表進行 Colorduino Shield 模組電路組立，再進行程式攥寫的動作(曹永忠, 許智诚, et al., 2014a; 曹永忠, 許智誠, et al., 2014a)。

表 46 Colorduino 接腳表

	模組接腳	Arduino 開發板接腳	解說
Colorduino	Vcc	itead arduino USB to TTL:Vcc	使用 USB2TTL 來燒錄 IDE 燒寫選擇 board 類型：Arduino Duemilanove w/atmega328
	GND	itead arduino USB to TTL:Gnd	
	Tx	itead arduino USB to TTL:Rx	
	Rx	itead arduino USB to TTL:Tx	

完成 Colorduino Shield 模組電路組立之後，將下表之 Colorduino 測試程式一，請讀者鍵入 S k e t c h　 I D E 軟體(軟體下載請到：https://www.arduino.cc/en/Main/Software)，編譯完成後上傳到開發版進行測試。

如下圖所示，可以見到炫麗的彩色在 RGB Led 燈板上不斷的變化，讀者也可以到筆者 Youtube 網站 https://www.youtube.com/watch?v=J9a-msxujRI，觀看程式執行的影片。

表 47 Colorduino 測試程式一

```
Colorduino 測試程式一(ColorduinoPlasma)
#include <Colorduino.h>

typedef struct
{
   unsigned char r;
   unsigned char g;
   unsigned char b;
} ColorRGB;

//a color with 3 components: h, s and v
```

```
typedef struct
{
    unsigned char h;
    unsigned char s;
    unsigned char v;
} ColorHSV;

unsigned char plasma[ColorduinoScreenWidth][ColorduinoScreenHeight];
long paletteShift;


//Converts an HSV color to RGB color
void HSVtoRGB(void *vRGB, void *vHSV)
{
    float r, g, b, h, s, v; //this function works with floats between 0 and 1
    float f, p, q, t;
    int i;
    ColorRGB *colorRGB=(ColorRGB *)vRGB;
    ColorHSV *colorHSV=(ColorHSV *)vHSV;

    h = (float)(colorHSV->h / 256.0);
    s = (float)(colorHSV->s / 256.0);
    v = (float)(colorHSV->v / 256.0);

    //if saturation is 0, the color is a shade of grey
    if(s == 0.0) {
        b = v;
        g = b;
        r = g;
    }
    //if saturation > 0, more complex calculations are needed
    else
    {
        h *= 6.0; //to bring hue to a number between 0 and 6, better for the calcula-
tions
        i = (int)(floor(h)); //e.g. 2.7 becomes 2 and 3.01 becomes 3 or 4.9999 becomes
4
        f = h - i;//the fractional part of h
```

```
    p = (float)(v * (1.0 - s));
    q = (float)(v * (1.0 - (s * f)));
    t = (float)(v * (1.0 - (s * (1.0 - f))));

    switch(i)
    {
      case 0: r=v; g=t; b=p; break;
      case 1: r=q; g=v; b=p; break;
      case 2: r=p; g=v; b=t; break;
      case 3: r=p; g=q; b=v; break;
      case 4: r=t; g=p; b=v; break;
      case 5: r=v; g=p; b=q; break;
      default: r = g = b = 0; break;
    }
  }
  colorRGB->r = (int)(r * 255.0);
  colorRGB->g = (int)(g * 255.0);
  colorRGB->b = (int)(b * 255.0);
}

float
dist(float a, float b, float c, float d)
{
  return sqrt((c-a)*(c-a)+(d-b)*(d-b));
}


void
plasma_morph()
{
  unsigned char x,y;
  float value;
  ColorRGB colorRGB;
  ColorHSV colorHSV;

  for(y = 0; y < ColorduinoScreenHeight; y++)
    for(x = 0; x < ColorduinoScreenWidth; x++) {
```

```
        {
        value = sin(dist(x + paletteShift, y, 128.0, 128.0) / 8.0)
            + sin(dist(x, y, 64.0, 64.0) / 8.0)
            + sin(dist(x, y + paletteShift / 7, 192.0, 64) / 7.0)
            + sin(dist(x, y, 192.0, 100.0) / 8.0);
        colorHSV.h=(unsigned char)((value) * 128)&0xff;
        colorHSV.s=255;
        colorHSV.v=255;
        HSVtoRGB(&colorRGB, &colorHSV);

        Colorduino.SetPixel(x, y, colorRGB.r, colorRGB.g, colorRGB.b);
        }
    }
    paletteShift++;

    Colorduino.FlipPage(); // swap screen buffers to show it
}


/*********************************************************
Name: ColorFill
Function: Fill the frame with a color
Parameter:R: the value of RED.      Range:RED 0~255
            G: the value of GREEN. Range:RED 0~255
            B: the value of BLUE.    Range:RED 0~255
*********************************************************/
void ColorFill(unsigned char R,unsigned char G,unsigned char B)
{
    PixelRGB *p = Colorduino.GetPixel(0,0);
    for (unsigned char y=0;y<ColorduinoScreenWidth;y++) {
        for(unsigned char x=0;x<ColorduinoScreenHeight;x++) {
            p->r = R;
            p->g = G;
            p->b = B;
            p++;
        }
    }

    Colorduino.FlipPage();
```

```
}

void setup()
{
    Colorduino.Init(); // initialize the board

    // compensate for relative intensity differences in R/G/B brightness
    // array of 6-bit base values for RGB (0~63)
    // whiteBalVal[0]=red
    // whiteBalVal[1]=green
    // whiteBalVal[2]=blue
    unsigned char whiteBalVal[3] = {36,63,63}; // for LEDSEE 6x6cm round matrix
    Colorduino.SetWhiteBal(whiteBalVal);


    // start with morphing plasma, but allow going to color cycling if desired.
    paletteShift=128000;
    unsigned char bcolor;

    //generate the plasma once
    for(unsigned char y = 0; y < ColorduinoScreenHeight; y++)
        for(unsigned char x = 0; x < ColorduinoScreenWidth; x++)
        {
            //the plasma buffer is a sum of sines
            bcolor = (unsigned char)
            (
                    128.0 + (128.0 * sin(x*8.0 / 16.0))
                + 128.0 + (128.0 * sin(y*8.0 / 16.0))
            ) / 2;
            plasma[x][y] = bcolor;
        }

    // to adjust white balance you can uncomment this line
    // and comment out the plasma_morph() in loop()
    // and then experiment with whiteBalVal above
    // ColorFill(255,255,255);
}
```

Colorduino 測試程式一(ColorduinoPlasma)

``
void loop()
{
    plasma_morph();
}
``

圖 82 Colorduino 測試程式一執行畫面

使用者自行繪製

對於使用者，我們必需可以自行指定要繪制的點、線、行、面等，所以我們將程式進一步修正，將下表之使用者自行繪製測試程式一，請讀者鍵入ＳｋｅｔｃｈＩＤＥ軟體(軟體下載請到：https://www.arduino.cc/en/Main/Software)，編譯完成後上傳到開發版進行測試。

如下圖所示，可以繪出三列四行的彩色畫面(曹永忠, 許智诚, et al., 2014a; 曹永忠, 許智誠, et al., 2014a)。

表 48 使用者自行繪製測試程式一

使用者自行繪製測試程式一(colorduino01)
``` #include <EEPROM.h> #include <Colorduino.h> ```

```
typedef struct
{
    unsigned char r;
    unsigned char g;
    unsigned char b;
} ColorRGB;

//a color with 3 components: h, s and v
typedef struct
{
    unsigned char h;
    unsigned char s;
    unsigned char v;
} ColorHSV;

unsigned char plasma[ColorduinoScreenWidth][ColorduinoScreenHeight];
long paletteShift;
int addr = 0;


void setup()
{
    Serial.begin(9600);
    Colorduino.Init(); // initialize the board

    // compensate for relative intensity differences in R/G/B brightness
    // array of 6-bit base values for RGB (0~63)
    // whiteBalVal[0]=red
    // whiteBalVal[1]=green
    // whiteBalVal[2]=blue
    unsigned char whiteBalVal[3] = {36,63,63}; // for LEDSEE 6x6cm round matrix
    Colorduino.SetWhiteBal(whiteBalVal);


    // start with morphing plasma, but allow going to color cycling if desired.
    paletteShift=128000;
    unsigned char bcolor;
```

```
//generate the plasma once
for(unsigned char y = 0; y < ColorduinoScreenHeight; y++)
    for(unsigned char x = 0; x < ColorduinoScreenWidth; x++)
    {
        //the plasma buffer is a sum of sines
        bcolor = (unsigned char)
        (
                128.0 + (128.0 * sin(x*8.0 / 16.0))
            + 128.0 + (128.0 * sin(y*8.0 / 16.0))
        ) / 2;
        plasma[x][y] = bcolor;
    }

// to adjust white balance you can uncomment this line
// and comment out the plasma_morph() in loop()
// and then experiment with whiteBalVal above
// ColorFill(255,255,255);
}

void loop()
{
    DrawLine(2,12,128,228);
    DrawLine(4,34,12,128);
    DrawLine(6,134,135,8);
    DrawColumn(0,132,128,228) ;
    DrawColumn(2,12,128,228) ;
    DrawColumn(4,34,12,128) ;
    DrawColumn(6,134,135,8) ;
    delay(1000);
}
//Converts an HSV color to RGB color
void HSVtoRGB(void *vRGB, void *vHSV)
{
    float r, g, b, h, s, v; //this function works with floats between 0 and 1
    float f, p, q, t;
    int i;
    ColorRGB *colorRGB=(ColorRGB *)vRGB;
```

```
ColorHSV *colorHSV=(ColorHSV *)vHSV;

h = (float)(colorHSV->h / 256.0);
s = (float)(colorHSV->s / 256.0);
v = (float)(colorHSV->v / 256.0);

//if saturation is 0, the color is a shade of grey
if(s == 0.0) {
    b = v;
    g = b;
    r = g;
}
//if saturation > 0, more complex calculations are needed
else
{
    h *= 6.0; //to bring hue to a number between 0 and 6, better for the calcula-
tions
    i = (int)(floor(h)); //e.g. 2.7 becomes 2 and 3.01 becomes 3 or 4.9999 becomes
4
    f = h - i;//the fractional part of h

    p = (float)(v * (1.0 - s));
    q = (float)(v * (1.0 - (s * f)));
    t = (float)(v * (1.0 - (s * (1.0 - f))));

    switch(i)
    {
        case 0: r=v; g=t; b=p; break;
        case 1: r=q; g=v; b=p; break;
        case 2: r=p; g=v; b=t; break;
        case 3: r=p; g=q; b=v; break;
        case 4: r=t; g=p; b=v; break;
        case 5: r=v; g=p; b=q; break;
        default: r = g = b = 0; break;
    }
}
colorRGB->r = (int)(r * 255.0);
colorRGB->g = (int)(g * 255.0);
```

```
    colorRGB->b = (int)(b * 255.0);
}


float
dist(float a, float b, float c, float d)
{
  return sqrt((c-a)*(c-a)+(d-b)*(d-b));
}



void plasma_morph()
{
   unsigned char x,y;
   float value;
   ColorRGB colorRGB;
   ColorHSV colorHSV;

   for(y = 0; y < ColorduinoScreenHeight; y++)
     for(x = 0; x < ColorduinoScreenWidth; x++) {
       {
     value = sin(dist(x + paletteShift, y, 128.0, 128.0) / 8.0)
        + sin(dist(x, y, 64.0, 64.0) / 8.0)
        + sin(dist(x, y + paletteShift / 7, 192.0, 64) / 7.0)
        + sin(dist(x, y, 192.0, 100.0) / 8.0);
     colorHSV.h=(unsigned char)((value) * 128)&0xff;
     colorHSV.s=255;
     colorHSV.v=255;
     HSVtoRGB(&colorRGB, &colorHSV);

     Colorduino.SetPixel(x, y, colorRGB.r, colorRGB.g, colorRGB.b);
       }
   }
   paletteShift++;

   Colorduino.FlipPage(); // swap screen buffers to show it
}

/*********************************************************
```

```
Name: ColorFill
Function: Fill the frame with a color
Parameter:R: the value of RED.      Range:RED 0~255
           G: the value of GREEN. Range:RED 0~255
           B: the value of BLUE.    Range:RED 0~255
************************************************************/
void ColorFill(unsigned char R,unsigned char G,unsigned char B)
{
  PixelRGB *p = Colorduino.GetPixel(0,0);
  for (unsigned char y=0;y<ColorduinoScreenWidth;y++) {
    for(unsigned char x=0;x<ColorduinoScreenHeight;x++) {
      p->r = R;
      p->g = G;
      p->b = B;
      p++;
    }
  }

  Colorduino.FlipPage();
}

void Draw(int x, int y, unsigned char R,unsigned char G,unsigned char B)
{
  PixelRGB *p = Colorduino.GetPixel(x,y);
    p->r = R;
    p->g = G;
    p->b = B;

    Colorduino.FlipPage();
}

void DrawLine(int line, unsigned char R,unsigned char G,unsigned char B)
{
  PixelRGB *p = Colorduino.GetPixel(0,line);
    for(unsigned char x=0;x<ColorduinoScreenHeight;x++) {
      p->r = R;
      p->g = G;
      p->b = B;
```

```
使用者自行繪製測試程式一(colorduino01)

        p++;
    }

  Colorduino.FlipPage();
}


void DrawColumn(int column, unsigned char R,unsigned char G,unsigned char B)
{
  for (unsigned char y=0;y<ColorduinoScreenWidth;y++) {
      PixelRGB *p = Colorduino.GetPixel(column,y);
      p->r = R;
      p->g = G;
      p->b = B;
      p++;
  }

  Colorduino.FlipPage();
}
```

圖 83 使用者自行繪製測試程式一執行畫面

隨機炫麗畫面的彩色字幕機

在來我們進一步需要加強功能，我們透過隨機變數來產生隨機顏色，透過隨機

顏色來繪出隨機彩色畫面，將下表之使用者自行繪製測試程式二，請讀者鍵入Ｓｋ
ｅｔｃｈ　ＩＤＥ軟體(軟體下載請到：https://www.arduino.cc/en/Main/Software)，
編譯完成後上傳到開發版進行測試。

　　如下圖所示，可以繪出隨機彩色畫面(曹永忠, 許智诚, et al., 2014a; 曹永忠, 許
智誠, et al., 2014a)。

表 49 使用者自行繪製測試程式二

使用者自行繪製測試程式二(colorduino02)
| ```
#include <EEPROM.h>
#include <Colorduino.h>

typedef struct
{
    unsigned char r;
    unsigned char g;
    unsigned char b;
} ColorRGB;

//a color with 3 components: h, s and v
typedef struct
{
    unsigned char h;
    unsigned char s;
    unsigned char v;
} ColorHSV;

unsigned char plasma[ColorduinoScreenWidth][ColorduinoScreenHeight];
long paletteShift;
int addr = 0;


void setup()
{
    randomSeed(millis());
``` |
```

```
    Serial.begin(9600);
    Colorduino.Init(); // initialize the board


    // compensate for relative intensity differences in R/G/B brightness
    // array of 6-bit base values for RGB (0~63)
    // whiteBalVal[0]=red
    // whiteBalVal[1]=green
    // whiteBalVal[2]=blue
    unsigned char whiteBalVal[3] = {36,63,63}; // for LEDSEE 6x6cm round matrix
    Colorduino.SetWhiteBal(whiteBalVal);


    // start with morphing plasma, but allow going to color cycling if desired.
    paletteShift=128000;
    unsigned char bcolor;

    //generate the plasma once
    for(unsigned char y = 0; y < ColorduinoScreenHeight; y++)
       for(unsigned char x = 0; x < ColorduinoScreenWidth; x++)
      {
         //the plasma buffer is a sum of sines
         bcolor = (unsigned char)
         (
                 128.0 + (128.0 * sin(x*8.0 / 16.0))
             + 128.0 + (128.0 * sin(y*8.0 / 16.0))
         ) / 2;
         plasma[x][y] = bcolor;
      }

   // to adjust white balance you can uncomment this line
   // and comment out the plasma_morph() in loop()
   // and then experiment with whiteBalVal above
   // ColorFill(255,255,255);
}

void loop()
{
  RamdomRGBFill();
```

```
    delay(1000);
}
//Converts an HSV color to RGB color
void HSVtoRGB(void *vRGB, void *vHSV)
{
    float r, g, b, h, s, v; //this function works with floats between 0 and 1
    float f, p, q, t;
    int i;
    ColorRGB *colorRGB=(ColorRGB *)vRGB;
    ColorHSV *colorHSV=(ColorHSV *)vHSV;

    h = (float)(colorHSV->h / 256.0);
    s = (float)(colorHSV->s / 256.0);
    v = (float)(colorHSV->v / 256.0);

    //if saturation is 0, the color is a shade of grey
    if(s == 0.0) {
        b = v;
        g = b;
        r = g;
    }
    //if saturation > 0, more complex calculations are needed
    else
    {
        h *= 6.0; //to bring hue to a number between 0 and 6, better for the calcula-
tions
        i = (int)(floor(h)); //e.g. 2.7 becomes 2 and 3.01 becomes 3 or 4.9999 becomes
4
        f = h - i;//the fractional part of h

        p = (float)(v * (1.0 - s));
        q = (float)(v * (1.0 - (s * f)));
        t = (float)(v * (1.0 - (s * (1.0 - f))));

        switch(i)
        {
            case 0: r=v; g=t; b=p; break;
            case 1: r=q; g=v; b=p; break;
```

```
          case 2: r=p; g=v; b=t; break;
          case 3: r=p; g=q; b=v; break;
          case 4: r=t; g=p; b=v; break;
          case 5: r=v; g=p; b=q; break;
          default: r = g = b = 0; break;
      }
  }
  colorRGB->r = (int)(r * 255.0);
  colorRGB->g = (int)(g * 255.0);
  colorRGB->b = (int)(b * 255.0);
}


float
dist(float a, float b, float c, float d)
{
  return sqrt((c-a)*(c-a)+(d-b)*(d-b));
}


void plasma_morph()
{
  unsigned char x,y;
  float value;
  ColorRGB colorRGB;
  ColorHSV colorHSV;

  for(y = 0; y < ColorduinoScreenHeight; y++)
    for(x = 0; x < ColorduinoScreenWidth; x++) {
      {
      value = sin(dist(x + paletteShift, y, 128.0, 128.0) / 8.0)
        + sin(dist(x, y, 64.0, 64.0) / 8.0)
        + sin(dist(x, y + paletteShift / 7, 192.0, 64) / 7.0)
        + sin(dist(x, y, 192.0, 100.0) / 8.0);
      colorHSV.h=(unsigned char)((value) * 128)&0xff;
      colorHSV.s=255;
      colorHSV.v=255;
      HSVtoRGB(&colorRGB, &colorHSV);
```

```
      Colorduino.SetPixel(x, y, colorRGB.r, colorRGB.g, colorRGB.b);
        }
    }
  paletteShift++;

  Colorduino.FlipPage(); // swap screen buffers to show it
}


/*********************************************************
Name: ColorFill
Function: Fill the frame with a color
Parameter:R: the value of RED.      Range:RED 0~255
            G: the value of GREEN. Range:RED 0~255
            B: the value of BLUE.    Range:RED 0~255
**********************************************************/
void ColorFill(unsigned char R,unsigned char G,unsigned char B)
{
  PixelRGB *p = Colorduino.GetPixel(0,0);
  for (unsigned char y=0;y<ColorduinoScreenWidth;y++) {
    for(unsigned char x=0;x<ColorduinoScreenHeight;x++) {
      p->r = R;
      p->g = G;
      p->b = B;
      p++;
    }
  }

  Colorduino.FlipPage();
}

void Draw(int x, int y, unsigned char R,unsigned char G,unsigned char B)
{
  PixelRGB *p = Colorduino.GetPixel(x,y);
      p->r = R;
      p->g = G;
      p->b = B;

    Colorduino.FlipPage();
```

```
}

void DrawLine(int line, unsigned char R,unsigned char G,unsigned char B)
{
    PixelRGB *p = Colorduino.GetPixel(0,line);
        for(unsigned char x=0;x<ColorduinoScreenHeight;x++) {
            p->r = R;
            p->g = G;
            p->b = B;
            p++;
        }

    Colorduino.FlipPage();
}

void DrawColumn(int column, unsigned char R,unsigned char G,unsigned char B)
{
    for (unsigned char y=0;y<ColorduinoScreenWidth;y++) {
        PixelRGB *p = Colorduino.GetPixel(column,y);
        p->r = R;
        p->g = G;
        p->b = B;
        p++;
    }

    Colorduino.FlipPage();
}
void RamdomRGBFill()
{
    PixelRGB *p = Colorduino.GetPixel(0,0);
    for (unsigned char y=0;y<ColorduinoScreenWidth;y++) {
        for(unsigned char x=0;x<ColorduinoScreenHeight;x++) {
            RamdomRGB(p) ;
            p++;
        }
    }

    Colorduino.FlipPage();
```

使用者自行繪製測試程式二(colorduino02)
| ```
}
void    RamdomRGB(PixelRGB *p)
{
        p->r = (byte)random(0,255);
         p->g = (byte)random(0,255);
         p->b = (byte)random(0,255);


}
``` |

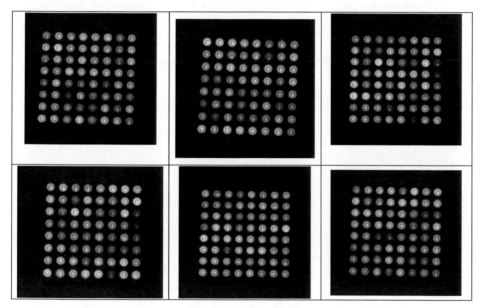

圖 84 使用者自行繪製測試程式二執行畫面

章節小結

　　本章主要介紹之 Arduino 開發板連接、使用彩色矩陣式 LED 顯示模組
(Colorduino Shield 模組)，透過本章節的解說，相信讀者會對連接、驅動彩色矩陣式
LED 顯示模組，有更深入的了解與體認。
```

CHAPTER

OLED 顯示螢幕

本章主要介紹常用於穿戴式裝置：如健康智慧手環、隨身 3C 產品：如 MP3 隨身聽…等等商業產品常用的顯示螢幕：OLED LCD，主要這些產品需要省電、高亮度、方便、顏色色彩多，高解析等特性，所以本文介紹 OLED LCD 顯示模組(曹永忠, 2016d)。

Oled 顯示器 I²C 版

目前有許多企業開發出許多彩色螢幕顯示器，本文主要介紹 OLED 顯示器[13]如下圖所示，由於 OLED 顯示器大同小異，大小與解析度不同以外，其原理大致相同，所以我們就介紹這款『OLED 顯示幕 0.96 寸 12864 IIC』為主要介紹主題，硬體部分讀者可以到電子商場、網路商場、電子材料商家，露天賣場、雅虎賣場、淘寶網等購買，本文主要是向 ICSHOP 賣場購買，購買參考網址如下：http://www.icshop.com.tw/product_info.php/products_id/22376，這是一款省電、高解析度，小型、高可視率的一款模組。

---

[13] OLED：Organic Light Emitting Display，即有機發光顯示器，在手機 LCD 上屬於新崛起的種類，被譽為 "夢幻顯示器"。OLED 的驅動方式可分為被動式矩陣（passive matrix，即 PM-OLED）與主動式矩陣（active matrix，即 AM-OLED）兩類，其中被動式矩陣架構較簡單，成本也較低，但必需在高脈衝電流下操作，才能達到適合人眼觀賞的亮度，因 OLED 的亮度與所通過的電流密度成正比，太高的操作電流不但會使電路效率及壽命降低，因為掃瞄的關係使其解析度也受限制，因此 PM-OLED 比較適合於小尺寸的產品。相反的，AM-OLED 雖然成本較昂貴、製程較複雜（仍比 TFT-LCD 容易），但其每一個畫素（pixel）皆可記憶驅動信號並可獨立與連續驅動，且效率較高，適用於大尺寸與高解析度之高資訊容量的顯示產品。

圖 85 Oled LCD I²C 板

如上圖所示，其實 OLED 顯示器有 SPI 介面與 I2C 介面，兩種模組大同小異試，共用相同的函數庫，本文使用的函數庫，為 Rinky-Dink Electronics 所分享的，其 官 網 為 ： http://www.rinkydinkelectronics.com/ ， 可 到 下 載 區 ： http://www.rinkydinkelectronics.com/library.php，下載，其 OLED 顯示幕 0.96 寸 12864 IIC 的函式庫網址為：http://www.rinkydinkelectronics.com/library.php?id=80 本文為了節省腳位，採用 I2C 介面的 OLED 顯示幕進行實驗(曹永忠, 2016d)。

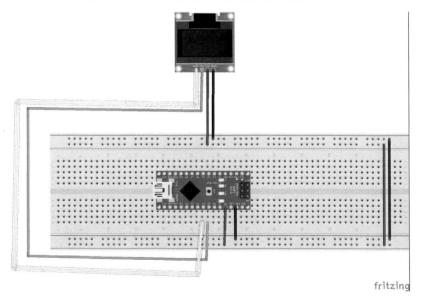

圖 86　OLED 顯示器連接電路圖

讀者可以參考上圖所示之 OLED 顯示器連接電路圖，對於 I2C 之腳位不太明瞭的，也可以參考下表之腳位說明，進行電路組立。

表 50 接腳表

接腳	接腳說明	接腳名稱
1	GND	接地 (0V) Arduino GND
2	VCC	電源 (+5V) Arduino +5V
3	SDA	Arduino SDA 腳位
4	SCL	Arduino SCL 腳位

如下表所示，為 OLED 顯示器測試程式，請讀者鍵入Ｓｋｅｔｃｈ ＩＤＥ軟體(軟體下載請到：https://www.arduino.cc/en/Main/Software)，編譯完成後上傳到開發版進行測試。

表 51 OLED 顯示器測試程式

OLED 顯示器測試程式(OLED_I2C_Graph_Demo)

// OLED_I2C_Graph_Demo
// Copyright (C)2015 Rinky-Dink Electronics, Henning Karlsen. All right reserved
// web: http://www.RinkyDinkElectronics.com/
//
// A quick demo of how to use my OLED_I2C library.
//
// To use the hardware I2C (TWI) interface of the Arduino you must connect
// the pins as follows:
//
// Arduino Uno/2009:

```
// --------------------
// Display:   SDA pin    -> Arduino Analog 4 or the dedicated SDA pin
//            SCL pin    -> Arduino Analog 5 or the dedicated SCL pin
//
// Arduino Leonardo:
// ---------------------
// Display:   SDA pin    -> Arduino Digital 2 or the dedicated SDA pin
//            SCL pin    -> Arduino Digital 3 or the dedicated SCL pin
//
// Arduino Mega:
// ---------------------
// Display:   SDA pin    -> Arduino Digital 20 (SDA) or the dedicated SDA pin
//            SCL pin    -> Arduino Digital 21 (SCL) or the dedicated SCL pin
//
// Arduino Due:
// ---------------------
// Display:   SDA pin    -> Arduino Digital 20 (SDA) or the dedicated SDA1
(Digital 70) pin
//            SCL pin    -> Arduino Digital 21 (SCL) or the dedicated SCL1
(Digital 71) pin
//
// The internal pull-up resistors will be activated when using the
// hardware I2C interfaces.
//
// You can connect the OLED display to any available pin but if you use
// any other than what is described above the library will fall back to
// a software-based, TWI-like protocol which will require exclusive access
// to the pins used, and you will also have to use appropriate, external
// pull-up resistors on the data and clock signals.
//

#include <OLED_I2C.h>

OLED   myOLED(SDA, SCL, 8);

extern uint8_t SmallFont[];
extern uint8_t logo[];
extern uint8_t The_End[];
extern uint8_t pacman1[];
```

```
extern uint8_t pacman2[];
extern uint8_t pacman3[];
extern uint8_t pill[];

float y;
uint8_t* bm;
int pacy;

void setup()
{
  myOLED.begin();
  myOLED.setFont(SmallFont);
  randomSeed(analogRead(7));
}

void loop()
{
  myOLED.clrScr();
  myOLED.drawBitmap(0, 16, logo, 128, 36);
  myOLED.update();

  delay(3000);

  myOLED.clrScr();
  myOLED.print("OLED_I2C", CENTER, 0);
  myOLED.print("DEMO", CENTER, 28);
  myOLED.drawRect(50, 26, 78, 36);
  for (int i=0; i<6; i++)
  {
    myOLED.drawLine(79, 26+(i*2), 105-(i*3), 26+(i*2));
    myOLED.drawLine(22+(i*3), 36-(i*2), 50, 36-(i*2));
  }
  myOLED.print("(C)2015 by", CENTER, 48);
  myOLED.print("Henning Karlsen", CENTER, 56);
  myOLED.update();

  delay(5000);

  myOLED.clrScr();
```

```
for (int i=0; i<64; i+=2)
{
  myOLED.drawLine(0, i, 127, 63-i);
  myOLED.update();
}
for (int i=127; i>=0; i-=2)
{
  myOLED.drawLine(i, 0, 127-i, 63);
  myOLED.update();
}

delay(2000);

myOLED.clrScr();
myOLED.drawRect(0, 0, 127, 63);
for (int i=0; i<64; i+=4)
{
  myOLED.drawLine(0, i, i*2, 63);
  myOLED.update();
}
for (int i=0; i<64; i+=4)
{
  myOLED.drawLine(127, 63-i, 127-(i*2), 0);
  myOLED.update();
}

delay(2000);

myOLED.clrScr();
for (int i=0; i<10; i++)
{
  myOLED.drawRoundRect(i*3, i*3, 127-(i*3), 63-(i*3));
  myOLED.update();
}

delay(2000);

myOLED.clrScr();
for (int i=0; i<25; i++)
```

```
{
  myOLED.drawCircle(64, 32, i*3);
  myOLED.update();
}

delay(2000);

myOLED.clrScr();
myOLED.drawRect(0, 0, 127, 63);
myOLED.drawLine(0, 31, 127, 31);
myOLED.drawLine(63, 0, 63, 63);
for (int c=0; c<4; c++)
{
  for (int i=0; i<128; i++)
  {
    y=i*0.049741883681838392942432518690191;
    myOLED.invPixel(i, (sin(y)*28)+31);
    myOLED.update();
    delay(10);
  }
}

delay(2000);

for (int pc=0; pc<3; pc++)
{
  pacy=random(0, 44);

  for (int i=-20; i<132; i++)
  {
    myOLED.clrScr();
    for (int p=6; p>((i+20)/20); p--)
      myOLED.drawBitmap(p*20-8, pacy+7, pill, 5, 5);
    switch(((i+20)/3) % 4)
    {
      case 0: bm=pacman1;
              break;
      case 1: bm=pacman2;
              break;
```

```
        case 2: bm=pacman3;
                break;
        case 3: bm=pacman2;
                break;
    }
    myOLED.drawBitmap(i, pacy, bm, 20, 20);
    myOLED.update();
    delay(10);
  }
}

for (int i=0; i<41; i++)
{
  myOLED.clrScr();
  myOLED.drawBitmap(22, i-24, The_End, 84, 24);
  myOLED.update();
  delay(50);
}
myOLED.print("Runtime (ms):", CENTER, 48);
myOLED.printNumI(millis(), CENTER, 56);
myOLED.update();
for (int i=0; i<5; i++)
{
  myOLED.invert(true);
  delay(1000);
  myOLED.invert(false);
  delay(1000);
}
}
```

參考網址：Rinky-Dink Electronics：http://www.rinkydinkelectronics.com/library.php

表 52 OLED 顯示器測試程式(Include File)

OLED 顯示器測試程式(graphics.c)
#include <avr/pgmspace.h>
const uint8_t logo[] PROGMEM={

```
    0x00, 0x00, 0x00, 0x00, 0x00, 0xF8, 0x08, 0x08, 0x08, 0x08, 0x08, 0x0F, 0x08,
0x08, 0x08, 0x08,     // 0x0010 (16) pixels
    0x08, 0xF8, 0x00, 0x00, 0x00, 0x00, 0x00, 0x00, 0x00, 0x00, 0x02, 0x03, 0x1F,
0xFF, 0xFE, 0x83,     // 0x0020 (32) pixels
    0x83, 0xC2, 0xC6, 0x7E, 0x3C, 0x1C, 0x00, 0x00, 0x00, 0x20, 0xE7, 0xE7, 0x00,
0x00, 0x00, 0x00,     // 0x0030 (48) pixels
    0x20, 0x70, 0xE0, 0xC0, 0x20, 0x30, 0x30, 0xE0, 0xC0, 0x00, 0x00, 0x03, 0x03,
0xFE, 0xFE, 0x00,     // 0x0040 (64) pixels
    0x20, 0xE0, 0x60, 0x20, 0x20, 0x00, 0x00, 0x40, 0xE0, 0xE0, 0x40, 0x00, 0x40,
0xE0, 0x60, 0x00,     // 0x0050 (80) pixels
    0x00, 0x00, 0x00, 0x00, 0x02, 0x02, 0xFF, 0xFF, 0xEF, 0x03, 0x02, 0x02, 0x02,
0x06, 0xFC, 0xF8,     // 0x0060 (96) pixels
    0xE0, 0x00, 0x00, 0x00, 0x20, 0xE7, 0xE7, 0x00, 0x00, 0x00, 0x20, 0x70, 0xE0,
0xC0, 0x20, 0x30,     // 0x0070 (112) pixels
    0x30, 0xE0, 0xC0, 0x00, 0x00, 0x00, 0x03, 0x03, 0xFE, 0xFE, 0x00, 0x20, 0xE0,
0x60, 0x20, 0x20,     // 0x0080 (128) pixels
    0x00, 0x00, 0xFF, 0x01, 0x01, 0x01, 0x01, 0x01, 0xFF, 0x00, 0x00, 0x00, 0x10,
0x10, 0x10, 0x10,     // 0x0090 (144) pixels
    0x90, 0x7F, 0x90, 0x10, 0x10, 0x10, 0x10, 0x00, 0x00, 0x00, 0x20, 0x20, 0x3C,
0x3F, 0x3F, 0x20,     // 0x00A0 (160) pixels
    0x00, 0x00, 0x07, 0x3F, 0x3C, 0x30, 0x30, 0x00, 0x00, 0x20, 0x3F, 0x3F, 0x30,
0x00, 0x00, 0x00,     // 0x00B0 (176) pixels
    0x20, 0x3F, 0x3F, 0x20, 0x00, 0x00, 0x20, 0x3F, 0x3F, 0x20, 0x00, 0x20, 0x20,
0x3F, 0x3F, 0x22,     // 0x00C0 (192) pixels
    0x27, 0x2F, 0x38, 0x30, 0x20, 0x00, 0x00, 0x00, 0x01, 0x0F, 0xBC, 0x60, 0x1E,
0x01, 0x00, 0x01,     // 0x00D0 (208) pixels
    0x01, 0x01, 0x01, 0x01, 0x20, 0x20, 0x3F, 0x3F, 0x3F, 0x20, 0x20, 0x20, 0x20,
0x30, 0x1F, 0x0F,     // 0x00E0 (224) pixels
    0x03, 0x00, 0x00, 0x00, 0x20, 0x3F, 0x3F, 0x30, 0x00, 0x00, 0x20, 0x3F, 0x3F,
0x20, 0x00, 0x00,     // 0x00F0 (240) pixels
    0x20, 0x3F, 0x3F, 0x20, 0x00, 0x00, 0x20, 0x20, 0x3F, 0x3F, 0x22, 0x27, 0x2F,
0x38, 0x30, 0x20,     // 0x0100 (256) pixels
    0x00, 0x00, 0xFF, 0x00, 0x00, 0x00, 0x00, 0x00, 0xFF, 0x00, 0x00, 0x00, 0x80,
0x60, 0x18, 0x06,     // 0x0110 (272) pixels
    0x01, 0x00, 0x01, 0x06, 0x18, 0x60, 0x80, 0x00, 0x00, 0x00, 0x00, 0x00, 0x00,
0x40, 0x40, 0xE0,     // 0x0120 (288) pixels
    0xE0, 0xC0, 0x40, 0x40, 0x40, 0x40, 0xC0, 0xC0, 0x00, 0x00, 0x40, 0xC0, 0xC0,
0x00, 0x00, 0x00,     // 0x0130 (304) pixels
    0x00, 0x00, 0x00, 0x00, 0x00, 0x00, 0x00, 0x00, 0x00, 0x00, 0x00, 0x00, 0x00,
```

```
0x00, 0x00, 0x00,      // 0x0140 (320) pixels
    0x00, 0x00, 0x00, 0xC0, 0x00, 0x00, 0x00, 0x03, 0x03, 0x02, 0x03, 0x00, 0x00,
0x00, 0x00, 0x00,      // 0x0150 (336) pixels
    0x00, 0x00, 0x00, 0x00, 0x00, 0x00, 0x00, 0x00, 0x00, 0x00, 0x00, 0x00, 0x00,
0x00, 0x00, 0x00,      // 0x0160 (352) pixels
    0x00, 0x00, 0x00, 0x00, 0x00, 0x00, 0x00, 0x00, 0xE0, 0xE0, 0x00, 0x00, 0x00,
0x00, 0x00, 0x00,      // 0x0170 (368) pixels
    0x00, 0x00, 0x00, 0x00, 0x00, 0x00, 0x00, 0x00, 0x00, 0x00, 0x00, 0x00, 0x00,
0x00, 0x00, 0x00,      // 0x0180 (384) pixels
    0x00, 0x00, 0x0F, 0x08, 0x08, 0xF8, 0x08, 0x08, 0x0F, 0x00, 0x00, 0x00, 0x01,
0x01, 0x01, 0x01,      // 0x0190 (400) pixels
    0x01, 0xFF, 0x01, 0x01, 0x01, 0x01, 0x01, 0x00, 0x00, 0x00, 0x00, 0x00, 0x00,
0x00, 0x00, 0xFF,      // 0x01A0 (416) pixels
    0xFF, 0x9F, 0x18, 0x18, 0x3E, 0x00, 0x00, 0x87, 0xC0, 0x00, 0x00, 0xFF, 0xFF,
0x00, 0x00, 0x00,      // 0x01B0 (432) pixels
    0x60, 0xF8, 0xFC, 0x46, 0x26, 0x2E, 0x3C, 0xB8, 0x00, 0xE0, 0xF8, 0x08, 0x0C,
0x0C, 0x7C, 0x38,      // 0x01C0 (448) pixels
    0x00, 0x06, 0x06, 0xFF, 0x0E, 0x06, 0x84, 0x00, 0x00, 0x04, 0xFC, 0xFC, 0x08,
0x0C, 0x0C, 0x08,      // 0x01D0 (464) pixels
    0x00, 0xF0, 0xF8, 0x1C, 0x06, 0x02, 0x04, 0xFC, 0xF8, 0xF0, 0x00, 0x04, 0xEE,
0xFC, 0x18, 0x04,      // 0x01E0 (480) pixels
    0x06, 0x06, 0xFC, 0xF8, 0x00, 0x00, 0x00, 0x04, 0xFC, 0xFC, 0x00, 0x00, 0x00,
0xE0, 0xF8, 0x08,      // 0x01F0 (496) pixels
    0x0C, 0x0C, 0x7C, 0x38, 0x00, 0x00, 0x7C, 0x7C, 0x64, 0xC4, 0xDC, 0xCC,
0x80, 0x00, 0x00, 0x00,      // 0x0200 (512) pixels
    0xF0, 0xF0, 0xF0, 0xF0, 0xF0, 0xF1, 0xF1, 0xF1, 0xF1, 0xF1, 0xF1, 0xFF, 0xF1,
0xF1, 0xF1, 0xF1,      // 0x0210 (528) pixels
    0xF1, 0xF1, 0xF0, 0xF0, 0xF0, 0xF0, 0xF0, 0xF0, 0xF0, 0xF0, 0xF0, 0xF0, 0xF0,
0xF4, 0xF4, 0xF7,      // 0x0220 (544) pixels
    0xF7, 0xF7, 0xF4, 0xF4, 0xF4, 0xF6, 0xF6, 0xF7, 0xF3, 0xF0, 0xF4, 0xF7, 0xF7,
0xF4, 0xF0, 0xF0,      // 0x0230 (560) pixels
    0xF0, 0xF3, 0xF7, 0xF4, 0xF4, 0xF4, 0xF6, 0xF1, 0xF0, 0xF1, 0xF7, 0xF6, 0xFC,
0xFC, 0xF4, 0xF7,      // 0x0240 (576) pixels
    0xF0, 0xF0, 0xF0, 0xF7, 0xFE, 0xF4, 0xF7, 0xF0, 0xF0, 0xF6, 0xF7, 0xF7, 0xF4,
0xF0, 0xF0, 0xF0,      // 0x0250 (592) pixels
    0xF0, 0xF0, 0xF3, 0xF6, 0xF4, 0xF4, 0xF4, 0xF7, 0xF3, 0xF0, 0xF0, 0xF4, 0xF7,
0xF7, 0xF4, 0xF0,      // 0x0260 (608) pixels
    0xF0, 0xF4, 0xF7, 0xF7, 0xF4, 0xF0, 0xF0, 0xF4, 0xF7, 0xF7, 0xF6, 0xF0, 0xF0,
0xF1, 0xF7, 0xF6,      // 0x0270 (624) pixels
```

```
    0xFC, 0xFC, 0xF4, 0xF7, 0xF0, 0xF0, 0xFF, 0xF6, 0xF4, 0xF8, 0xF8, 0xF7, 0xF1,
0xF0, 0xF0, 0xF0,     // 0x0280 (640) pixels
    };


    const uint8_t The_End[] PROGMEM={
    0x00, 0x80, 0x80, 0xC0, 0xC0, 0xC0, 0xC0, 0xC0, 0xC0, 0xC0, 0xC0, 0xC0, 0xC0,
0xE0, 0x60, 0x00, 0x00,     // 0x0010 (16) pixels
    0x80, 0xC0, 0xC0, 0x00, 0x00, 0xC0, 0xC0, 0xC0, 0x00, 0x00, 0x80, 0xC0, 0xC0,
0xC0, 0xC0, 0xE0,     // 0x0020 (32) pixels
    0xC0, 0xC0, 0xC0, 0xC0, 0xE0, 0x60, 0x00, 0x00, 0x00, 0x00, 0x00, 0xC0, 0xC0,
0xC0, 0xC0, 0xE0,     // 0x0030 (48) pixels
    0xE0, 0xC0, 0xC0, 0xC0, 0xE0, 0xE0, 0x00, 0x00, 0x00, 0x80, 0x80, 0xC0, 0xC0,
0xC0, 0x00, 0x00,     // 0x0040 (64) pixels
    0x00, 0x00, 0x80, 0xE0, 0xF0, 0xF0, 0x60, 0x40, 0xF0, 0xF0, 0xF0, 0xF0, 0xE0,
0xE0, 0xC0, 0xC0,     // 0x0050 (80) pixels
    0x80, 0x00, 0x00, 0x00, 0x00, 0x03, 0x03, 0x03, 0x81, 0xFC, 0xFF, 0x0F, 0x03,
0x00, 0x00, 0x00,     // 0x0060 (96) pixels
    0xEE, 0x6F, 0x67, 0xFF, 0xFF, 0x7F, 0x71, 0x30, 0xF0, 0xFF, 0x3F, 0x39, 0x38,
0x18, 0x00, 0x01,     // 0x0070 (112) pixels
    0x00, 0xF8, 0xFF, 0x1F, 0x0F, 0x0C, 0x0D, 0x8D, 0x80, 0x80, 0x00, 0x00, 0x00,
0x00, 0x00, 0x01,     // 0x0080 (128) pixels
    0x00, 0xC0, 0xFF, 0x7F, 0x0F, 0x0C, 0x0D, 0x0D, 0x84, 0x80, 0x80, 0x07, 0x07,
0x83, 0xFF, 0xFF,     // 0x0090 (144) pixels
    0x1F, 0x3F, 0xFE, 0xF8, 0xF8, 0xFE, 0xDF, 0x03, 0x00, 0x00, 0x00, 0x03, 0xFF,
0xFF, 0x1F, 0x00,     // 0x00A0 (160) pixels
    0x00, 0x80, 0x81, 0xC3, 0xE7, 0x7F, 0x3E, 0x00, 0x00, 0x00, 0x00, 0x06, 0x07,
0x07, 0x03, 0x00,     // 0x00B0 (176) pixels
    0x00, 0x00, 0x00, 0x00, 0x00, 0x00, 0x04, 0x0F, 0x07, 0x03, 0x00, 0x00, 0x0F,
0x0F, 0x07, 0x00,     // 0x00C0 (192) pixels
    0x00, 0x06, 0x06, 0x3F, 0x1F, 0x0F, 0x0F, 0x0E, 0x06, 0x06, 0x06, 0x07, 0x07,
0x03, 0x00, 0x00,     // 0x00D0 (208) pixels
    0x00, 0x00, 0x06, 0x1F, 0x3F, 0x1F, 0x0F, 0x0E, 0x06, 0x06, 0x06, 0x07, 0x07,
0x07, 0x03, 0x00,     // 0x00E0 (224) pixels
    0x06, 0x0F, 0x07, 0x01, 0x00, 0x00, 0x00, 0x01, 0x0F, 0x07, 0x03, 0x00, 0x00,
0x18, 0x1C, 0x1F,     // 0x00F0 (240) pixels
    0x0F, 0x07, 0x06, 0x07, 0x03, 0x03, 0x01, 0x01, 0x00, 0x00, 0x00, 0x00,
    };


    const uint8_t pacman1[] PROGMEM={
```

```
    0x80, 0xE0, 0xF0, 0xF8, 0xFC, 0xFE, 0xFE, 0xFF, 0xFF, 0xFF, 0xFF, 0xFF, 0xFF,
0x7E, 0x3E, 0x1C,     // 0x0010 (16) pixels
    0x0C, 0x00, 0x00, 0x00, 0x1F, 0x7F, 0xFF, 0xFF, 0xFF, 0xFF, 0xFF, 0xFF, 0xFF,
0xFF, 0xFF, 0xF9,     // 0x0020 (32) pixels
    0xF0, 0xE0, 0xC0, 0x80, 0x00, 0x00, 0x00, 0x00, 0x00, 0x00, 0x00, 0x01, 0x03,
0x07, 0x07, 0x0F,     // 0x0030 (48) pixels
    0x0F, 0x0F, 0x0F, 0x0F, 0x0F, 0x07, 0x07, 0x03, 0x03, 0x00, 0x00, 0x00,
    };


    const uint8_t pacman2[] PROGMEM={
    0x80, 0xE0, 0xF0, 0xF8, 0xFC, 0xFE, 0xFE, 0xFF, 0xFF, 0xFF, 0xFF, 0xFF, 0xFF,
0xFE, 0xFE, 0x7C,     // 0x0010 (16) pixels
    0x7C, 0x38, 0x20, 0x00, 0x1F, 0x7F, 0xFF, 0xFF, 0xFF, 0xFF, 0xFF, 0xFF, 0xFF,
0xFF, 0xFF, 0xF9,     // 0x0020 (32) pixels
    0xF9, 0xF0, 0xF0, 0xE0, 0xE0, 0xC0, 0x40, 0x00, 0x00, 0x00, 0x00, 0x01, 0x03,
0x07, 0x07, 0x0F,     // 0x0030 (48) pixels
    0x0F, 0x0F, 0x0F, 0x0F, 0x0F, 0x07, 0x07, 0x03, 0x03, 0x01, 0x00, 0x00,
    };


    const uint8_t pacman3[] PROGMEM={
    0x80, 0xE0, 0xF0, 0xF8, 0xFC, 0xFE, 0xFE, 0xFF, 0xFF, 0xFF, 0xFF, 0xFF, 0xFF,
0xFE, 0xFE, 0xFC,     // 0x0010 (16) pixels
    0xF8, 0xF0, 0xE0, 0x80, 0x1F, 0x7F, 0xFF, 0xFF, 0xFF, 0xFF, 0xFF, 0xFF, 0xFF,
0xFF, 0xFF, 0xFF,     // 0x0020 (32) pixels
    0xFF, 0xFF, 0xFF, 0xFF, 0xFB, 0xF9, 0x79, 0x19, 0x00, 0x00, 0x00, 0x01, 0x03,
0x07, 0x07, 0x0F,     // 0x0030 (48) pixels
    0x0F, 0x0F, 0x0F, 0x0F, 0x0F, 0x07, 0x07, 0x03, 0x01, 0x00, 0x00, 0x00,
    };


    const uint8_t pill[] PROGMEM={
    0x0E, 0x1F, 0x1F, 0x1F, 0x0E,
    };
```

程式下載網址：https://github.com/brucetsao/makerdiwo/tree/master/201606

參考網址：Rinky-Dink Electronics：http://www.rinkydinkelectronics.com/library.php

讀　者　也　可　以　在　筆　者　　YouTube　　頻　　道

(https://www.youtube.com/user/UltimaBruce) 中 ， 在 網 址
https://www.youtube.com/watch?v=L7WMQKdNlUU&feature=youtu.be，看到本次實驗-
OLED 顯示幕測試程式結果畫面。

　　如下圖所示，我們可以看到 Arduino 在 OLED 顯示幕畫面上顯示圖片情形。

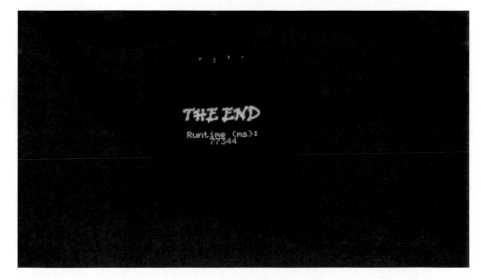

圖 87 OLED 顯示幕測試程式結果畫面

章節小結

　　本章主要介紹之 Arduino 開發板使用與連接 OLED 顯示模組，透過本章節的解
說，相信讀者會對連接、使用 OLED 顯示模組，有更深入的了解與體認。

10

CHAPTER

2.4~3.2" TFT 顯示模組

隨著數位化技術的推進，電子數位應用產品為人類生活帶來極佳的便利性與樂趣，而所有數位產品皆需要顯示介面來呈現內容，因此，顯示模組已成為產業的關注焦點，不同的顯模組技術在其中角逐，全球相關廠商也投注資源開發顯模組技之新技術、新應用，並為提昇人類視覺享受而努力。其中，TFT-LCD(Thin Film Transistor Liquid Crystal Display--薄膜電晶體液晶顯示器)產品具有輕、薄、省能源、低幅射之優點，已被視為主流顯示技術(曹永忠, 2016b, 2016c)。

　　本篇主要介紹常用於隨身裝置：如健康智慧手環、隨身 3C 產品：如 MP3 隨身聽…等等商業產品常用的顯示螢幕(曹永忠, 2016d)：薄膜電晶體液晶顯示器(Thin film transistor liquid crystal display：TFT-LCD)，主要的特性是高亮度、操控簡單方便、顏色色彩多，高解析、體積小等特性，所以本文介紹薄膜電晶體液晶顯示模組(Thin film transistor liquid crystal display：TFT-LCD)。

薄膜電晶體液晶顯示器

　　薄膜電晶體液晶顯示器[14]（英語：Thin film transistor liquid crystal display，一般簡稱為 TFT-LCD）是多數液晶顯示器的一種，它使用薄膜電晶體技術改善影像品質。雖然 TFT-LCD 被統稱為 LCD，不過它是種主動式矩陣 LCD，被應用在電視、平面顯示器及投影機上(曹永忠, 2016b)。

　　簡單說，TFT-LCD 面板可視為兩片玻璃基板中間夾著一層液晶，上層的玻璃基板是與彩色濾光片、而下層的玻璃則有電晶體鑲嵌於上。當電流通過電晶體產生電場變化，造成液晶分子偏轉，藉以改變光線的偏極性，再利用偏光片決定畫素的明暗狀態。上層玻璃因與彩色濾光片貼合，形成每個畫素各包含紅藍綠三顏色，這

---

[14]薄膜電晶體液晶顯示器(Wiki)
https://zh.wikipedia.org/wiki/%E8%96%84%E8%86%9C%E9%9B%BB%E6%99%B6%E9%AB%94%E6%B6%B2%E6%99%B6%E9%A1%AF%E7%A4%BA%E5%99%A8

些發出紅藍綠色彩光線的畫素，便形成了面板上的影像畫面(曹永忠, 2016b)。

薄膜電晶體-液晶顯示器(TFT-LCD)的面板基本架構如下圖所示：

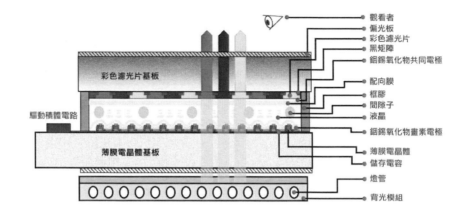

圖 88 薄膜電晶體-液晶顯示器(TFT-LCD)的面板基本架構圖

參考資料：http://sbh770803.blogspot.tw/2013_03_01_archive.html、

http://kunya-sh.com/jishu/20141205/14.html

簡單的講，它是一個以電信號控制的光開關裝置。液晶介於兩片透明導電之鉬錫氧化物(ITO)電極之間，經由加在鉬錫氧化物(ITO)電極上的電壓高低可以控制不同的液晶排列方向(如下圖所示)，而液晶的排列方向與光線的穿透量有關，進而造成畫素的亮暗程度不同，這就是灰階的控制原理(顏色則是由彩色濾光片產生)。此畫素的灰階是由資料驅動器(Data driver)所能提供的分電壓數目決定。

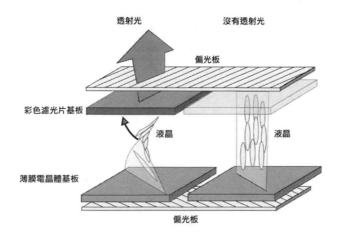

圖 89 液晶基本光電特性

參考資料：http://sbh770803.blogspot.tw/2013_03_01_archive.html、

http://kunya-sh.com/jishu/20141205/14.html

圖 90 TFT 2.4” 板

如上圖所示，其實薄膜電晶體液晶顯示模組(Thin film transistor liquid crystal display：TFT-LCD)驅動方式非常簡單，下列為常見 TFT 2.4” 板的規格(曹永忠，2016b)：

TFT 2.4” 板的規格

螢幕尺寸 ： 2.4 inch

解析度 ： 240 x 320

色彩數 r ： 65k
驅動方法 ：SPI
控制介面 ：8080 8 data bit with 4 control bits
觸控面板 ：4 線電阻式觸控面板

如上述所示，下列為常見 TFT 2.4"板常見接腳表：

表 53 TFT 2.4"板常見接腳表

Arduino 腳位	TFT 2.4 擴充版腳位	用途
3.3V	3.3V	Power
5V	5V	Power
GND	GND	Power
A0	LCD_RD	LCD Control
A1	LCD_WR TOUCH_YP	LCD Control / Touch Data
A2	LCD_RS TOUCH_XM	LCD Control / Touch Data
A3	LCD_CS	LCD Control
A4	LCD_RST	LCD Reset
D2	LCD_D2	LCD Data
D3	LCD_D3	LCD Data
D4	LCD_D4	LCD Data
D5	LCD_D5	LCD Data
D6	LCD_D6 / TOUCH XP	LCD Data/ Touch Data
D7	LCD_D7 / TOUCH YM	LCD Data / Touch Data
D8	LCD_D0	LCD Data
D9	LCD_D1	LCD Data
D10	SD_CS	SD Select
D11	SD_DI	SD Data
D12	SD_DO	SD Data
D13	SD_SCK	SD Clock

筆者乃是透過 ICSHOP 官網購買 2.4 "& 2.8" TFT 顯示模組，網址如下：

http://www.icshop.com.tw/product_info.php/products_id/12997，也可以到淘寶網：都會

明武電子(https://shop111496966.world.taobao.com/?spm=a312a.7700824.0.0.UA9SlE)，2.4

寸 TFT 液晶屏觸控式螢幕彩屏模組，網址如下：

https://world.taobao.com/item/39834819734.htm?fromSite=main&spm=a312a.7700846.0.0.H

iBw40&_u=tvlvti9a9c2，本文範例乃採用此塊 2.4 "TFT 顯示模組，進行介紹。

　　讀者可以參考下圖，Adafruit 公司官網：Adafruit 2.4" TFT LCD with Touchscreen

Breakout w/MicroSD Socket ，網址：https://www.adafruit.com/product/2478，with

Touchscreen Breakout w/MicroSD Socket 示之 2.4 "TFT 顯示模組連接電路圖，對於 I2C

之腳位不太明瞭的，也可以參考下表之腳位說明，進行電路組立。

　　讀者可以 Adafruit 公司官網：Adafruit 2.4" Color TFT Touchscreen Breakout，網

址： https://cdn-learn.adafruit.com/downloads/pdf/adafruit-2-4-color-tft-touchscreen-breakout.pdf，

裡面有更詳細的介紹，筆者就在這裡不多敘述，有興趣的讀者，可以自行參閱該網

址的文件檔。

圖 91　24 "TFT 顯示器連接電路圖

參考資料：Adafruit 2.4" TFT LCD with Touchscreen Breakout w/MicroSD Socket -

ILI9341(https://learn.adafruit.com/assets/25676)

讀者可以參考上圖所示之〝TFT 顯示器連接電路圖,也可以參考下表之腳位說明,進行電路組立。

表 54 〝TFT 顯示器接腳表

如下表所示,為 2.4〝TFT 顯示模組,請讀者鍵入Ｓｋｅｔｃｈ　ＩＤＥ軟體(軟體下載請到:https://www.arduino.cc/en/Main/Software),編譯完成後上傳到開發版進行測試。

本程式需要用到兩個函式庫(曹永忠, 2016g),關於 Adafruit_TFTLCD.h,請到網址:https://github.com/brucetsao/LIB_for_MCU/tree/master/Arduino_Lib/libraries/TFTLCD,進行下載,關於 Adafruit_GFX.h,請到網址:

https://github.com/brucetsao/LIB_for_MCU/tree/master/Arduino_Lib/libraries/Adafruit

_GFX,進行下載(曹永忠, 2016b)。

表 55 24 〝TFT 顯示模組測試程式一

24 〝TFT 顯示模組測試程式一(MCU_OK_graphicstest)
// IMPORTANT: Adafruit_TFTLCD LIBRARY MUST BE SPECIFICALLY // CONFIGURED FOR EITHER THE TFT SHIELD OR THE BREAKOUT BOARD. // SEE RELEVANT COMMENTS IN Adafruit_TFTLCD.h FOR SETUP.

```
#include <Adafruit_GFX.h>      // Core graphics library
#include <Adafruit_TFTLCD.h> // Hardware-specific library

// The control pins for the LCD can be assigned to any digital or
// analog pins...but we'll use the analog pins as this allows us to
// double up the pins with the touch screen (see the TFT paint example).
#define LCD_CS A3 // Chip Select goes to Analog 3
#define LCD_CD A2 // Command/Data goes to Analog 2
#define LCD_WR A1 // LCD Write goes to Analog 1
#define LCD_RD A0 // LCD Read goes to Analog 0

#define LCD_RESET A4 // Can alternately just connect to Arduino's reset pin

// When using the BREAKOUT BOARD only, use these 8 data lines to the LCD:
// For the Arduino Uno, Duemilanove, Diecimila, etc.:
//     D0 connects to digital pin 8    (Notice these are
//     D1 connects to digital pin 9     NOT in order!)
//     D2 connects to digital pin 2
//     D3 connects to digital pin 3
//     D4 connects to digital pin 4
//     D5 connects to digital pin 5
//     D6 connects to digital pin 6
//     D7 connects to digital pin 7
// For the Arduino Mega, use digital pins 22 through 29
// (on the 2-row header at the end of the board).

// Assign human-readable names to some common 16-bit color values:
#define   BLACK     0x0000
#define BLUE        0x001F
#define RED         0xF800
#define GREEN       0x07E0
#define CYAN        0x07FF
#define MAGENTA 0xF81F
#define YELLOW    0xFFE0
#define WHITE      0xFFFF

Adafruit_TFTLCD tft(LCD_CS, LCD_CD, LCD_WR, LCD_RD, LCD_RESET);
// If using the shield, all control and data lines are fixed, and
```

```
    // a simpler declaration can optionally be used:
    // Adafruit_TFTLCD tft;

    void setup(void) {
      Serial.begin(9600);
      Serial.println(F("TFT LCD test"));

    #ifdef USE_ADAFRUIT_SHIELD_PINOUT
      Serial.println(F("Using Adafruit 2.8\" TFT Arduino Shield Pinout"));
    #else
      Serial.println(F("Using Adafruit 2.8\" TFT Breakout Board Pinout"));
    #endif

      Serial.print("TFT size is "); Serial.print(tft.width()); Serial.print("x");
Serial.println(tft.height());

      tft.reset();

      uint16_t identifier = tft.readID();

      if(identifier == 0x9325) {
        Serial.println(F("Found ILI9325 LCD driver"));
      } else if(identifier == 0x9327) {
        Serial.println(F("Found ILI9327 LCD driver"));
      } else if(identifier == 0x9328) {
        Serial.println(F("Found ILI9328 LCD driver"));
      } else if(identifier == 0x7575) {
        Serial.println(F("Found HX8347G LCD driver"));
      } else if(identifier == 0x9341) {
        Serial.println(F("Found ILI9341 LCD driver"));
      } else if(identifier == 0x8357) {
        Serial.println(F("Found HX8357D LCD driver"));
      } else if(identifier == 0x0154) {
        Serial.println(F("Found S6D0154 LCD driver"));
      } else {
        Serial.print(F("Unknown LCD driver chip: "));
        Serial.println(identifier, HEX);
        Serial.println(F("If using the Adafruit 2.8\" TFT Arduino shield, the line:"));
        Serial.println(F("  #define USE_ADAFRUIT_SHIELD_PINOUT"));
```

```
      Serial.println(F("should appear in the library header (Adafruit_TFT.h)."));
      Serial.println(F("If using the breakout board, it should NOT be #defined!"));
      Serial.println(F("Also if using the breakout, double-check that all wiring"));
      Serial.println(F("matches the tutorial."));
      return;
   }

   tft.begin(identifier);

   Serial.println(F("Benchmark                    Time (microseconds)"));

   Serial.print(F("Screen fill                 "));
   Serial.println(testFillScreen());
   delay(500);

   Serial.print(F("Text                        "));
   Serial.println(testText());
   delay(3000);

   Serial.print(F("Lines                       "));
   Serial.println(testLines(CYAN));
   delay(500);

   Serial.print(F("Horiz/Vert Lines            "));
   Serial.println(testFastLines(RED, BLUE));
   delay(500);

   Serial.print(F("Rectangles (outline)        "));
   Serial.println(testRects(GREEN));
   delay(500);

   Serial.print(F("Rectangles (filled)         "));
   Serial.println(testFilledRects(YELLOW, MAGENTA));
   delay(500);

   Serial.print(F("Circles (filled)            "));
   Serial.println(testFilledCircles(10, MAGENTA));

   Serial.print(F("Circles (outline)           "));
```

```
    Serial.println(testCircles(10, WHITE));
    delay(500);

    Serial.print(F("Triangles (outline)        "));
    Serial.println(testTriangles());
    delay(500);

    Serial.print(F("Triangles (filled)         "));
    Serial.println(testFilledTriangles());
    delay(500);

    Serial.print(F("Rounded rects (outline)   "));
    Serial.println(testRoundRects());
    delay(500);

    Serial.print(F("Rounded rects (filled)    "));
    Serial.println(testFilledRoundRects());
    delay(500);

    Serial.println(F("Done!"));
}

void loop(void) {
    for(uint8_t rotation=0; rotation<4; rotation++) {
        tft.setRotation(rotation);
        testText();
        delay(2000);
    }
}

unsigned long testFillScreen() {
    unsigned long start = micros();
    tft.fillScreen(BLACK);
    tft.fillScreen(RED);
    tft.fillScreen(GREEN);
    tft.fillScreen(BLUE);
    tft.fillScreen(BLACK);
    return micros() - start;
}
```

```
unsigned long testText() {
  tft.fillScreen(BLACK);
  unsigned long start = micros();
  tft.setCursor(0, 0);
  tft.setTextColor(WHITE);    tft.setTextSize(1);
  tft.println("Hello World!");
  tft.setTextColor(YELLOW); tft.setTextSize(2);
  tft.println(1234.56);
  tft.setTextColor(RED);        tft.setTextSize(3);
  tft.println(0xDEADBEEF, HEX);
  tft.println();
  tft.setTextColor(GREEN);
  tft.setTextSize(5);
  tft.println("Groop");
  tft.setTextSize(2);
  tft.println("I implore thee,");
  tft.setTextSize(1);
  tft.println("my foonting turlingdromes.");
  tft.println("And hooptiously drangle me");
  tft.println("with crinkly bindlewurdles,");
  tft.println("Or I will rend thee");
  tft.println("in the gobberwarts");
  tft.println("with my blurglecruncheon,");
  tft.println("see if I don't!");
  return micros() - start;
}


unsigned long testLines(uint16_t color) {
  unsigned long start, t;
  int              x1, y1, x2, y2,
                   w = tft.width(),
                   h = tft.height();

  tft.fillScreen(BLACK);

  x1 = y1 = 0;
  y2    = h - 1;
  start = micros();
```

```
for(x2=0; x2<w; x2+=6) tft.drawLine(x1, y1, x2, y2, color);
x2     = w - 1;
for(y2=0; y2<h; y2+=6) tft.drawLine(x1, y1, x2, y2, color);
t      = micros() - start; // fillScreen doesn't count against timing

tft.fillScreen(BLACK);

x1     = w - 1;
y1     = 0;
y2     = h - 1;
start = micros();
for(x2=0; x2<w; x2+=6) tft.drawLine(x1, y1, x2, y2, color);
x2     = 0;
for(y2=0; y2<h; y2+=6) tft.drawLine(x1, y1, x2, y2, color);
t      += micros() - start;

tft.fillScreen(BLACK);

x1     = 0;
y1     = h - 1;
y2     = 0;
start = micros();
for(x2=0; x2<w; x2+=6) tft.drawLine(x1, y1, x2, y2, color);
x2     = w - 1;
for(y2=0; y2<h; y2+=6) tft.drawLine(x1, y1, x2, y2, color);
t      += micros() - start;

tft.fillScreen(BLACK);

x1     = w - 1;
y1     = h - 1;
y2     = 0;
start = micros();
for(x2=0; x2<w; x2+=6) tft.drawLine(x1, y1, x2, y2, color);
x2     = 0;
for(y2=0; y2<h; y2+=6) tft.drawLine(x1, y1, x2, y2, color);

return micros() - start;
}
```

```
unsigned long testFastLines(uint16_t color1, uint16_t color2) {
    unsigned long start;
    int          x, y, w = tft.width(), h = tft.height();

    tft.fillScreen(BLACK);
    start = micros();
    for(y=0; y<h; y+=5) tft.drawFastHLine(0, y, w, color1);
    for(x=0; x<w; x+=5) tft.drawFastVLine(x, 0, h, color2);

    return micros() - start;
}

unsigned long testRects(uint16_t color) {
    unsigned long start;
    int          n, i, i2,
                 cx = tft.width()   / 2,
                 cy = tft.height() / 2;

    tft.fillScreen(BLACK);
    n     = min(tft.width(), tft.height());
    start = micros();
    for(i=2; i<n; i+=6) {
        i2 = i / 2;
        tft.drawRect(cx-i2, cy-i2, i, i, color);
    }

    return micros() - start;
}

unsigned long testFilledRects(uint16_t color1, uint16_t color2) {
    unsigned long start, t = 0;
    int          n, i, i2,
                 cx = tft.width()   / 2 - 1,
                 cy = tft.height() / 2 - 1;

    tft.fillScreen(BLACK);
    n = min(tft.width(), tft.height());
    for(i=n; i>0; i-=6) {
```

```
        i2     = i / 2;
      start = micros();
      tft.fillRect(cx-i2, cy-i2, i, i, color1);
      t      += micros() - start;
      // Outlines are not included in timing results
      tft.drawRect(cx-i2, cy-i2, i, i, color2);
    }

    return t;
  }

unsigned long testFilledCircles(uint8_t radius, uint16_t color) {
    unsigned long start;
    int x, y, w = tft.width(), h = tft.height(), r2 = radius * 2;

    tft.fillScreen(BLACK);
    start = micros();
    for(x=radius; x<w; x+=r2) {
      for(y=radius; y<h; y+=r2) {
        tft.fillCircle(x, y, radius, color);
      }
    }

    return micros() - start;
  }

unsigned long testCircles(uint8_t radius, uint16_t color) {
    unsigned long start;
    int                 x, y, r2 = radius * 2,
                        w = tft.width()   + radius,
                        h = tft.height() + radius;

    // Screen is not cleared for this one -- this is
    // intentional and does not affect the reported time.
    start = micros();
    for(x=0; x<w; x+=r2) {
      for(y=0; y<h; y+=r2) {
        tft.drawCircle(x, y, radius, color);
      }
```

```
  }

    return micros() - start;
}

unsigned long testTriangles() {
  unsigned long start;
  int              n, i, cx = tft.width()    / 2 - 1,
                        cy = tft.height() / 2 - 1;

  tft.fillScreen(BLACK);
  n        = min(cx, cy);
  start = micros();
  for(i=0; i<n; i+=5) {
    tft.drawTriangle(
        cx        , cy - i, // peak
        cx - i, cy + i, // bottom left
        cx + i, cy + i, // bottom right
        tft.color565(0, 0, i));
  }

    return micros() - start;
}

unsigned long testFilledTriangles() {
  unsigned long start, t = 0;
  int              i, cx = tft.width()    / 2 - 1,
                        cy = tft.height() / 2 - 1;

  tft.fillScreen(BLACK);
  start = micros();
  for(i=min(cx,cy); i>10; i-=5) {
    start = micros();
    tft.fillTriangle(cx, cy - i, cx - i, cy + i, cx + i, cy + i,
      tft.color565(0, i, i));
    t += micros() - start;
    tft.drawTriangle(cx, cy - i, cx - i, cy + i, cx + i, cy + i,
      tft.color565(i, i, 0));
  }
```

```
    return t;
}

unsigned long testRoundRects() {
    unsigned long start;
    int          w, i, i2,
                 cx = tft.width()   / 2 - 1,
                 cy = tft.height() / 2 - 1;

    tft.fillScreen(BLACK);
    w      = min(tft.width(), tft.height());
    start = micros();
    for(i=0; i<w; i+=6) {
        i2 = i / 2;
        tft.drawRoundRect(cx-i2, cy-i2, i, i, i/8, tft.color565(i, 0, 0));
    }

    return micros() - start;
}

unsigned long testFilledRoundRects() {
    unsigned long start;
    int          i, i2,
                 cx = tft.width()   / 2 - 1,
                 cy = tft.height() / 2 - 1;

    tft.fillScreen(BLACK);
    start = micros();
    for(i=min(tft.width(), tft.height()); i>20; i-=6) {
        i2 = i / 2;
        tft.fillRoundRect(cx-i2, cy-i2, i, i, i/8, tft.color565(0, i, 0));
    }

    return micros() - start;
}
```

程式下載網址：https://github.com/brucetsao/makerdiwo/tree/master/201607

讀者也可以在筆者 YouTube 頻道

(https://www.youtube.com/user/UltimaBruce)中，在網址

https://www.youtube.com/watch?v=Raqd71_C_2Y&feature=youtu.be，看到本次實驗-

24"TFT 顯示模組測試程式一結果畫面。

如下圖所示，我們可以看到 Arduino 在 24"TFT 顯示模組畫面上顯示圖片情形。

圖 92 24 "TFT 顯示模組測試程式一結果畫面

顯示模組上 SD 卡讀寫模組

本段主要介紹讀者，如何使用 24 "TFT 顯示模組上的 SD Card 模組。

首先，因為我們要讀取圖片檔，所以我們先講解 SD Card 讀寫簡單原理，

讀者可以參考下圖之接腳方式，進行電路組立(曹永忠, 2016b, 2016c)。

圖 93 24 "TFTTFT 顯示器接腳圖

如下圖所示，為 2.4 "TFT 顯示模組進行讀寫裝置， 如下圖所示，將 Micro SD 卡插入下圖紅框處，該模組可以支援到 16G 容量的 Micro SD 卡。

圖 94   24 "TFT 顯示模組 SD 卡插槽處

我們請讀者鍵入 S k e t c h   I D E 軟體 (軟體下載請到：https://www.arduino.cc/en/Main/Software)，編譯完成後上傳到開發版進行測試。

表 56 24 "TFT 顯示模組 SD 卡測試程式一

24 "TFT 顯示模組 SD 卡測試程式一(MCU_OK_SD_Info)
/* SD card test This example shows how use the utility libraries on which the' SD library is based in order to get info about your SD card. Very useful for testing a card when you're not sure whether its working or not.

```
  The circuit:
   * SD card attached to SPI bus as follows:
  ** MOSI - pin 11 on Arduino Uno/Duemilanove/Diecimila
  ** MISO - pin 12 on Arduino Uno/Duemilanove/Diecimila
  ** CLK - pin 13 on Arduino Uno/Duemilanove/Diecimila
  ** CS - depends on your SD card shield or module.
             Pin 4 used here for consistency with other Arduino examples


  created    28 Mar 2011
  by Limor Fried
  modified 9 Apr 2012
  by Tom Igoe
  */
// include the SD library:
#include <SPI.h>
#include <SD.h>

// set up variables using the SD utility library functions:
Sd2Card card;
SdVolume volume;
SdFile root;

// change this to match your SD shield or module;
// Arduino Ethernet shield: pin 4
// Adafruit SD shields and modules: pin 10
// Sparkfun SD shield: pin 8
const int chipSelect = 10;

void setup() {
   // Open serial communications and wait for port to open:
   Serial.begin(9600);
   while (!Serial) {
      ; // wait for serial port to connect. Needed for native USB port only
   }


   Serial.print("\nInitializing SD card...");
```

```
// we'll use the initialization code from the utility libraries
// since we're just testing if the card is working!
if (!card.init(SPI_HALF_SPEED, chipSelect)) {
    Serial.println("initialization failed. Things to check:");
    Serial.println("* is a card inserted?");
    Serial.println("* is your wiring correct?");
    Serial.println("* did you change the chipSelect pin to match your shield or
module?");
    return;
} else {
    Serial.println("Wiring is correct and a card is present.");
}

// print the type of card
Serial.print("\nCard type: ");
switch (card.type()) {
    case SD_CARD_TYPE_SD1:
        Serial.println("SD1");
        break;
    case SD_CARD_TYPE_SD2:
        Serial.println("SD2");
        break;
    case SD_CARD_TYPE_SDHC:
        Serial.println("SDHC");
        break;
    default:
        Serial.println("Unknown");
}

// Now we will try to open the 'volume'/'partition' - it should be FAT16 or FAT32
if (!volume.init(card)) {
    Serial.println("Could not find FAT16/FAT32 partition.\nMake sure you've
formatted the card");
    return;
}

// print the type and size of the first FAT-type volume
```

```
        uint32_t volumesize;
        Serial.print("\nVolume type is FAT");
        Serial.println(volume.fatType(), DEC);
        Serial.println();

        volumesize = volume.blocksPerCluster();      // clusters are collections of blocks
        volumesize *= volume.clusterCount();          // we'll have a lot of clusters
        volumesize *= 512;                                    // SD card blocks are
always 512 bytes
        Serial.print("Volume size (bytes): ");
        Serial.println(volumesize);
        Serial.print("Volume size (Kbytes): ");
        volumesize /= 1024;
        Serial.println(volumesize);
        Serial.print("Volume size (Mbytes): ");
        volumesize /= 1024;
        Serial.println(volumesize);


        Serial.println("\nFiles found on the card (name, date and size in bytes): ");
        root.openRoot(volume);

        // list all files in the card with date and size
        root.ls(LS_R | LS_DATE | LS_SIZE);
    }



    void loop(void) {

    }
```

程式下載網址：https://github.com/brucetsao/makerdiwo/tree/master/201607

　　如下圖所示，我們可以看到 Arduino 在 24"TFT 顯示模組讀取 Micro SD 卡的
內容，將其 Micro SD 卡的資訊顯示出來(曹永忠, 2016b, 2016c; 曹永忠, 許智誠, et
al., 2015f, 2015g, 2015h, 2015i, 2015j, 2015l; 曹永忠, 許碩芳, et al., 2015a, 2015b)。

圖 95 24 "TFT 顯示模組 SD 卡測試程式一結果畫面

顯示模組顯示圖片

本段主要介紹讀者，如何將讀出的照片，顯示在 24 "TFT 顯示模組上。首先，我們參考 adafruit 官網之學習網站，網址如下：https://learn.adafruit.com/2-8-tft-touchscreen/bitmaps，先行將 woof.bmp，下載網址：https://github.com/brucetsao/makerdiwo/tree/master/201607/images，如下圖所示，先將圖片：woof.bmp 下載後，放到 Micro SD 卡的根目錄，因為我們要顯示該張圖片(曹永忠, 2016b, 2016c)。

圖 96 24 "TFT 顯示模組顯示圖片程式所用圖片

參考網址：https://learn.adafruit.com/2-8-tft-touchscreen/bitmaps

我們將下表程式，請讀者鍵入 Sketch IDE 軟體(軟體下載請到：

https://www.arduino.cc/en/Main/Software)，編譯完成後上傳到開發版進行測試。

表 57 24 "TFT 顯示模組顯示圖片程式一

24 "TFT 顯示模組顯示圖片程式一(MCU_OK_tftbmp)

// BMP-loading example specifically for the TFTLCD breakout board.
// If using the Arduino shield, use the tftbmp_shield.pde sketch instead!
// If using an Arduino Mega make sure to use its hardware SPI pins, OR make
// sure the SD library is configured for 'soft' SPI in the file Sd2Card.h.
 	
#include <Adafruit_GFX.h>   // Core graphics library
#include <Adafruit_TFTLCD.h> // Hardware-specific library
#include <SD.h>
#include <SPI.h>
 	
// The control pins for the LCD can be assigned to any digital or
// analog pins...but we'll use the analog pins as this allows us to
// double up the pins with the touch screen (see the TFT paint example).
#define LCD_CS A3 // Chip Select goes to Analog 3
#define LCD_CD A2 // Command/Data goes to Analog 2
#define LCD_WR A1 // LCD Write goes to Analog 1
#define LCD_RD A0 // LCD Read goes to Analog 0
 	
// When using the BREAKOUT BOARD only, use these 8 data lines to the LCD:
// For the Arduino Uno, Duemilanove, Diecimila, etc.:
//   D0 connects to digital pin 8   (Notice these are
//   D1 connects to digital pin 9   NOT in order!)
//   D2 connects to digital pin 2
//   D3 connects to digital pin 3
//   D4 connects to digital pin 4
//   D5 connects to digital pin 5
//   D6 connects to digital pin 6

```
//     D7 connects to digital pin 7
// For the Arduino Mega, use digital pins 22 through 29
// (on the 2-row header at the end of the board).

// For Arduino Uno/Duemilanove, etc
//    connect the SD card with DI going to pin 11, DO going to pin 12 and SCK
going to pin 13 (standard)
//    Then pin 10 goes to CS (or whatever you have set up)
#define SD_CS 10        // Set the chip select line to whatever you use (10 doesnt
conflict with the library)

// In the SD card, place 24 bit color BMP files (be sure they are 24-bit!)
// There are examples in the sketch folder

// our TFT wiring
Adafruit_TFTLCD tft(LCD_CS, LCD_CD, LCD_WR, LCD_RD, A4);

void setup()
{
    Serial.begin(9600);

    tft.reset();

    uint16_t identifier = tft.readID();

        if(identifier == 0x9325) {
        Serial.println(F("Found ILI9325 LCD driver"));
    } else if(identifier == 0x9327) {
        Serial.println(F("Found ILI9327 LCD driver"));
    } else if(identifier == 0x9328) {
        Serial.println(F("Found ILI9328 LCD driver"));
    } else if(identifier == 0x7575) {
        Serial.println(F("Found HX8347G LCD driver"));
    } else if(identifier == 0x9341) {
        Serial.println(F("Found ILI9341 LCD driver"));
    } else if(identifier == 0x8357) {
        Serial.println(F("Found HX8357D LCD driver"));
    } else if(identifier == 0x0154) {
        Serial.println(F("Found S6D0154 LCD driver"));
```

```
    } clsc {
      Serial.print(F("Unknown LCD driver chip: "));
      Serial.println(identifier, HEX);
      Serial.println(F("If using the Adafruit 2.8\" TFT Arduino shield, the line:"));
      Serial.println(F("   #define USE_ADAFRUIT_SHIELD_PINOUT"));
      Serial.println(F("should appear in the library header (Adafruit_TFT.h)."));
      Serial.println(F("If using the breakout board, it should NOT be #defined!"));
      Serial.println(F("Also if using the breakout, double-check that all wiring"));
      Serial.println(F("matches the tutorial."));
      return;
    }

    tft.begin(identifier);

    Serial.print(F("Initializing SD card..."));
    if (!SD.begin(SD_CS)) {
      Serial.println(F("failed!"));
      return;
    }
    Serial.println(F("OK!"));

    bmpDraw("woof.bmp", 0, 0);
    delay(1000);
}

void loop()
{
    for(int i = 0; i<4; i++) {
      tft.setRotation(i);
      tft.fillScreen(0);
      for(int j=0; j <= 200; j += 50) {
        bmpDraw("miniwoof.bmp", j, j);
      }
      delay(1000);
    }
}

// This function opens a Windows Bitmap (BMP) file and
// displays it at the given coordinates.   It's sped up
```

```
// by reading many pixels worth of data at a time
// (rather than pixel by pixel).   Increasing the buffer
// size takes more of the Arduino's precious RAM but
// makes loading a little faster.   20 pixels seems a
// good balance.

#define BUFFPIXEL 20

void bmpDraw(char *filename, int x, int y) {

    File        bmpFile;
    int         bmpWidth, bmpHeight;    // W+H in pixels
    uint8_t   bmpDepth;                          // Bit depth (currently must be 24)
    uint32_t bmpImageoffset;              // Start of image data in file
    uint32_t rowSize;                            // Not always = bmpWidth; may have
padding
    uint8_t   sdbuffer[3*BUFFPIXEL]; // pixel in buffer (R+G+B per pixel)
    uint16_t lcdbuffer[BUFFPIXEL];   // pixel out buffer (16-bit per pixel)
    uint8_t   buffidx = sizeof(sdbuffer); // Current position in sdbuffer
    boolean   goodBmp = false;         // Set to true on valid header parse
    boolean   flip       = true;          // BMP is stored bottom-to-top
    int         w, h, row, col;
    uint8_t   r, g, b;
    uint32_t pos = 0, startTime = millis();
    uint8_t   lcdidx = 0;
    boolean   first = true;

    if((x >= tft.width()) || (y >= tft.height())) return;

    Serial.println();
    Serial.print(F("Loading image '"));
    Serial.print(filename);
    Serial.println('\"');
    // Open requested file on SD card
    if ((bmpFile = SD.open(filename)) == NULL) {
        Serial.println(F("File not found"));
        return;
    }
```

```
// Parse BMP header
if(read16(bmpFile) == 0x4D42) { // BMP signature
  Serial.println(F("File size: ")); Serial.println(read32(bmpFile));
  (void)read32(bmpFile); // Read & ignore creator bytes
  bmpImageoffset = read32(bmpFile); // Start of image data
  Serial.print(F("Image Offset: ")); Serial.println(bmpImageoffset, DEC);
  // Read DIB header
  Serial.print(F("Header size: ")); Serial.println(read32(bmpFile));
  bmpWidth   = read32(bmpFile);
  bmpHeight = read32(bmpFile);
  if(read16(bmpFile) == 1) { // # planes -- must be '1'
    bmpDepth = read16(bmpFile); // bits per pixel
    Serial.print(F("Bit Depth: ")); Serial.println(bmpDepth);
    if((bmpDepth == 24) && (read32(bmpFile) == 0)) { // 0 = uncompressed

      goodBmp = true; // Supported BMP format -- proceed!
      Serial.print(F("Image size: "));
      Serial.print(bmpWidth);
      Serial.print('x');
      Serial.println(bmpHeight);

      // BMP rows are padded (if needed) to 4-byte boundary
      rowSize = (bmpWidth * 3 + 3) & ~3;

      // If bmpHeight is negative, image is in top-down order.
      // This is not canon but has been observed in the wild.
      if(bmpHeight < 0) {
        bmpHeight = -bmpHeight;
        flip      = false;
      }

      // Crop area to be loaded
      w = bmpWidth;
      h = bmpHeight;
      if((x+w-1) >= tft.width())   w = tft.width()   - x;
      if((y+h-1) >= tft.height()) h = tft.height() - y;

      // Set TFT address window to clipped image bounds
      tft.setAddrWindow(x, y, x+w-1, y+h-1);
```

```
for (row=0; row<h; row++) { // For each scanline...
    // Seek to start of scan line.    It might seem labor-
    // intensive to be doing this on every line, but this
    // method covers a lot of gritty details like cropping
    // and scanline padding.    Also, the seek only takes
    // place if the file position actually needs to change
    // (avoids a lot of cluster math in SD library).
    if(flip) // Bitmap is stored bottom-to-top order (normal BMP)
        pos = bmpImageoffset + (bmpHeight - 1 - row) * rowSize;
    else        // Bitmap is stored top-to-bottom
        pos = bmpImageoffset + row * rowSize;
    if(bmpFile.position() != pos) { // Need seek?
        bmpFile.seek(pos);
        buffidx = sizeof(sdbuffer); // Force buffer reload
    }

    for (col=0; col<w; col++) { // For each column...
        // Time to read more pixel data?
        if (buffidx >= sizeof(sdbuffer)) { // Indeed
            // Push LCD buffer to the display first
            if(lcdidx > 0) {
                tft.pushColors(lcdbuffer, lcdidx, first);
                lcdidx = 0;
                first   = false;
            }
            bmpFile.read(sdbuffer, sizeof(sdbuffer));
            buffidx = 0; // Set index to beginning
        }

        // Convert pixel from BMP to TFT format
        b = sdbuffer[buffidx++];
        g = sdbuffer[buffidx++];
        r = sdbuffer[buffidx++];
        lcdbuffer[lcdidx++] = tft.color565(r,g,b);
    } // end pixel
} // end scanline
// Write any remaining data to LCD
if(lcdidx > 0) {
```

```
            tft.pushColors(lcdbuffcr, lcdidx, first);
        }
        Serial.print(F("Loaded in "));
        Serial.print(millis() - startTime);
        Serial.println(" ms");
      } // end goodBmp
    }
  }

  bmpFile.close();
  if(!goodBmp) Serial.println(F("BMP format not recognized."));
}

// These read 16- and 32-bit types from the SD card file.
// BMP data is stored little-endian, Arduino is little-endian too.
// May need to reverse subscript order if porting elsewhere.

uint16_t read16(File f) {
  uint16_t result;
  ((uint8_t *)&result)[0] = f.read(); // LSB
  ((uint8_t *)&result)[1] = f.read(); // MSB
  return result;
}

uint32_t read32(File f) {
  uint32_t result;
  ((uint8_t *)&result)[0] = f.read(); // LSB
  ((uint8_t *)&result)[1] = f.read();
  ((uint8_t *)&result)[2] = f.read();
  ((uint8_t *)&result)[3] = f.read(); // MSB
  return result;
}
```

程式下載網址：https://github.com/brucetsao/makerdiwo/tree/master/201607

讀者也可以在筆者 YouTube 頻道(https://www.youtube.com/user/UltimaBruce)中，在
網址 https://www.youtube.com/watch?v=DhYMwefOJYM&feature=youtu.be，看到本次實
驗- 24 "TFT 顯示模組顯示圖片程式一結果畫面。

如下圖所示，我們可以看到 Arduino 在 24 "TFT 顯示模組上讀取 Micro SD 卡的圖片內容，將其該圖片顯示在 TFT 顯示模組上。

圖 97 24 "TFT 顯示模組顯示圖片程式一結果畫面

顯示模組上觸控模組

本段主要介紹讀者，如何使用 24 "TFT 顯示模組上的觸控功能。使用這個觸控功能，可以使用我們的指甲或觸控筆，基於這個部份，讀者只要使用電阻式的觸控筆，注意不可以使用電容式觸控筆，因為觸控原理不一樣，無法使用。

本程式需要用到 TouchScreen 函式庫，關於 TouchScreen.h，請到網址：https://github.com/brucetsao/LIB_for_MCU/tree/master/Arduino_Lib/libraries/TouchScreen 、https://github.com/adafruit/Touch-Screen-Library，進行下載與安裝(曹永忠, 2016a, 2016b, 2016c, 2016g)。

我們參考下表程式，請讀者鍵入Ｓｋｅｔｃｈ　ＩＤＥ軟體(軟體下載請到：https://www.arduino.cc/en/Main/Software)，編譯完成後上傳到開發版進行測試。

表 58 24 "TFT 顯示模組之觸控模組測試程式一

24 "TFT 顯示模組之觸控模組測試程式一(MCU_OK_touchscreendemo)
// Touch screen library with X Y and Z (pressure) readings as well // as oversampling to avoid 'bouncing'

```
// This demo code returns raw readings, public domain

#include <stdint.h>
#include "TouchScreen.h"

#define YP A2    // must be an analog pin, use "An" notation!
#define XM A3    // must be an analog pin, use "An" notation!
#define YM 8     // can be a digital pin
#define XP 9     // can be a digital pin

// For better pressure precision, we need to know the resistance
// between X+ and X- Use any multimeter to read it
// For the one we're using, its 300 ohms across the X plate
TouchScreen ts = TouchScreen(XP, YP, XM, YM, 300);

void setup(void) {
   Serial.begin(9600);
}

void loop(void) {
   // a point object holds x y and z coordinates
   TSPoint p = ts.getPoint();

   // we have some minimum pressure we consider 'valid'
   // pressure of 0 means no pressing!
   if (p.z > ts.pressureThreshhold) {
      Serial.print("X = "); Serial.print(p.x);
      Serial.print("\tY = "); Serial.print(p.y);
      Serial.print("\tPressure = "); Serial.println(p.z);
   }

   delay(100);
}
```

程式下載網址：https://github.com/brucetsao/makerdiwo/tree/master/201607

如下圖所示，我們可以看到 Arduino 在 24 〝TFT 顯示模組上使用觸控模組，進行讀取觸控座標，並將座標資訊顯示出來。

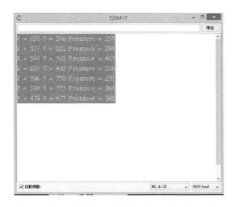

圖 98　24 "TFT 顯示模組之觸控模組測試程式一結果畫面

章節小結

　　本章主要介紹之 Arduino 開發板使用與連接 2.4~3.2" TFT 顯示模組，透過本章節的解說，相信讀者會對連接、使用 2.4~3.2" TFT 顯示模組，有更深入的了解與體認。

11

CHAPTER

NOKIA 5110 LCD 顯示模組

隨著數位化技術的推進，電子數位應用產品為人類生活帶來極佳的便利性與樂趣，而所有數位產品皆需要顯示介面來呈現內容，因此，顯示模組已成為產業的關注焦點，不同的顯模組技術在其中角逐，全球相關廠商也投注資源開發顯模組技之新技術、新應用，並為提昇人類視覺享受而努力

液晶顯示器（Liquid-Crystal Display：LCD）為平面薄型的顯示裝置，由一定數量的彩色或黑白畫素組成，放置於光源或者反射面前方。由於液晶顯示器功耗低，價格便宜，驅動容易，因此倍受工程師青睞，適用於使用電池的電子裝置。

本篇主要介紹常用於隨身裝置：如健康智慧手環、隨身 3C 產品：如 MP3 隨身聽...等等商業產品常用的顯示螢幕(曹永忠, 2016d)：液晶顯示器（Liquid-Crystal Display：LCD）主要的特性功耗低，價格便宜，驅動容易等特性，所以本文介紹 NOKIA 5110 圖形顯示的 LCD 螢幕，主要這款螢幕，被手機大廠 NOKIA 用來裝置在 5110 系列手機，因而大受歡迎(曹永忠, 2016e, 2016f)。

液晶顯示器（Liquid-Crystal Display：LCD）基本介紹

如下圖所示，液晶顯示器的每個畫素由以下幾個部分構成：懸浮於兩個透明電極（氧化銦錫）間的一列液晶分子層，兩邊外側有兩個偏振方向互相垂直的偏振過濾片。如果沒有電極間的液晶，光通過其中一個偏振過濾片其偏振方向將和第二個偏振片完全垂直，因此被完全阻擋了。但是如果通過一個偏振過濾片的光線偏振方向被液晶旋轉，那麼它就可以通過另一個偏振過濾片。液晶對光線偏振方向的旋轉可以通過靜電場控制，從而實作對光的控制。

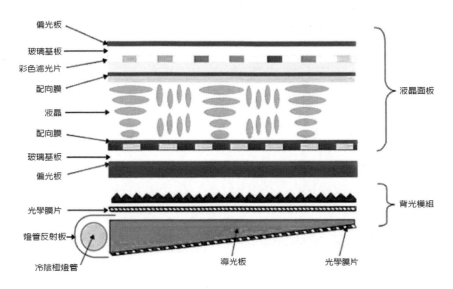

圖 99 液晶顯示器構造圖

資料來源：由 I, Wasami007，創用 CC 姓名標示-相同方式分享 3.0，

https://commons.wikimedia.org/w/index.php?curid=2430923

　　液晶分子極易受外加電場的影響而產生感應電荷。將少量的電荷加到每個畫素或者子畫素的透明電極產生靜電場，則液晶的分子將被此靜電場誘發感應電荷並產生靜電扭力，而使液晶分子原本的旋轉排列產生變化，因此也改變通過光線的旋轉幅度。改變一定的角度，從而能夠通過偏振過濾片。

　　在將電荷加到透明電極之前，液晶分子的排列被電極表面的排列決定，電極的化學物質表面可作為晶體的晶種。在最常見的扭曲向列型液晶　（Twisted Nematic Liquid Crystal：TN）中，液晶上下兩個電極垂直排列。液晶分子螺旋排列，通過一個偏振過濾片的光線在通過液晶片後偏振方向發生旋轉，從而能夠通過另一個偏振片。在此過程中一小部分光線被偏振片阻擋，從外面看上去是灰色。將電荷加到透明電極上後，液晶分子將幾乎完全順著電場方向平行排列，因此透過一個偏振過濾片的光線偏振方向沒有旋轉，因此光線被完全阻擋了。此時畫素看上去是黑色。通過控制電壓，可以控制液晶分子排列的扭曲程度，從而達到不同的灰度。

有些液晶顯示器在交流電作用下變黑，交流電破壞了液晶的螺旋效應，而關閉電流後，液晶顯示器會變亮或者透明，這類液晶顯示器常見於筆記型電腦與平價液晶顯示器上。另一類常應用於高畫質液晶顯示器或大型液晶電視上的液晶顯示器則是在關閉電源時，液晶顯示器為不透光的狀態。

　　為了省電，液晶顯示器採用復用的方法，在復用模式下，一端的電極分組連線在一起，每一組電極連線到一個電源，另一端的電極也分組連線，每一組連線到電源另一端，分組設計保證每個畫素由一個獨立的電源控制，電子裝置或者驅動電子裝置的軟體通過控制電源的開/關序列，從而控制畫素的顯示(資料來源：https://zh.wikipedia.org/wiki/%E6%B6%B2%E6%99%B6%E6%98%BE%E7%A4%BA%E5%99%A8)。

運作原理

　　在不加電壓下，光線會沿著液晶分子的間隙前進而轉折 90 度，所以光可通過。但加入電壓後，光順著液晶分子的間隙直線前進，因此光線會被濾光片所阻隔。

　　液晶是具有流動特性的物質，所以只需外加很微小的力量即可使液晶分子運動，以最常見普遍的向列型液晶為例，液晶分子可輕易的藉著電場作用使得液晶分子轉向，由於液晶的光軸與其分子軸相當一致，故可藉此產生光學效果，而當加於液晶的電場移除消失時，液晶將藉著其本身的彈性及黏性，液晶分子將十分迅速的回復原來未加電場前的狀態(資料來源：https://zh.wikipedia.org/wiki/%E6%B6%B2%E6%99%B6%E6%98%BE%E7%A4%BA%E5%99%A8)。

顯示方式

　　液晶顯示器可透射顯示，也可反射顯示，決定於它的光源放哪裡。一般來說，透射型液晶顯示器由一個螢幕背後的光源照亮，而觀看則在螢幕另一邊（前面）。

這種類型的 LCD 多用在需高亮度顯示的應用中,例如電腦顯示器、PDA 和手機中。用於照亮液晶顯示器的照明裝置的功耗往往高於液晶顯示器本身。

反射型液晶顯示器,常見於電子鐘錶和計算機中,(有時候)由後面的散射的反射面將外部的光反射回來照亮螢幕。這種類型的液晶顯示器具有較高的對比度,因為光線要經過液晶兩次,所以被削減了兩次。不使用照明裝置明顯降低了功耗,因此使用電池的裝置電池使用更久。因為小型的反射型液晶顯示器功耗非常低,以至於光電池就足以給它供電,因此常用於袖珍型計算機。

半穿透反射式液晶顯示器既可以當作透射型使用,也可當作反射型使用。當外部光線很足的時候,該液晶顯示器按照反射型工作,而當外部光線不足的時候,它又能當作透射型使用(資料來源: https://zh.wikipedia.org/wiki/%E6%B6%B2%E6%99%B6%E6%98%BE%E7%A4%BA%E5%99%A8)。

Nokia 5110 LCD 模組基本驅動方法

NOKIA 5110 是一款基於圖形顯示的 LCD 螢幕(如下圖所示),主要這款螢幕,被手機大廠 NOKIA 用來裝置在 5110 系列手機,因而大受歡迎並且有很多的應用。

Nokia 5110 LCD 使用的 PCD8544 控制器,NOKIA 3110 用的也是這款控制器,PCD8554 是一款低功耗的 CMOS LCD 控制器,用於驅動 48 行 84 列的圖形顯示,並且採用串列匯流排界面與微控制器相連,大大減少了週邊控制線的數量,在使用時十分的方便,而且相對於 LCD1602、LCD12864 都有著自己的獨特優勢(曹永忠, 2016e, 2016f)。

圖 100 Nokia 5110 LCD 模組

我們可以查詢 Sparkfun 官網：產品：Graphic LCD 84x48 - Nokia 5110，網址如下：https://www.sparkfun.com/products/10168，可以看到 Nokia 5110 LCD 模組的所有資料與規格。

Nokia 5110 LCD 模組規格如下：

- 單片 LCD 控制器

- 顯示尺寸：25 x 35mm

- 顯示資料緩衝區 48 x 84 位元，最多可顯示四行中文字

- 序列介面，最大速率 4Mbits/S

- 供電電壓 2.7V 到 3.3V，極限 7V

- 低功耗，可用電池供電

- 採用串行接點與主處理器進行通信，接點信號線數量大幅度減少，包括電源和地在內的信號線僅有 9 條。支援多種串行通信協議（如 AVR 單片機的ＳＰＩ、MCS51 的連接模式０等），傳輸速率高達 4Mbps，可全速寫入顯示數據，無等待時間

- 可通過導電膠連接模組與印製版，而不用連接電纜，用模塊上的金屬鉤可將模組固定到印製板上，因而非常便於安裝和更換

- LCD 控制器 / 驅動器芯片已綁定到 LCD 晶片上，模塊的體積很小

● 採用低電壓供電，正常顯示時的工作電流在 200μA 以下，且具有掉電模式

在 Arduino 官網：http://playground.arduino.cc/Code/PCD8544，也有介紹 Nokia 5110 LCD 模組，讀者可以參考下表之腳位說明，進行電路組立(曹永忠, 2016b, 2016c)。

表 59 "TFT 顯示器接腳表

Nokia 5110 LCD		開發版	
接腳	接腳說明	腳位	用途
VCC	5V(正)	+5C	供電(正)
GND	接地	GND	供電(接地)
SCE/CE	晶片致能	D7	數位腳位 7
RST/Reset	晶片重置	D6	數位腳位 6
D/C	Data/Command select 資料/命令選擇	D5	數位腳位 5
DN/MOSI	Serial Data Out 資料輸出	D4	數位腳位 4
SCLK	Serial Clock Out 脈波控制輸出	D3	數位腳位 3
LED	被光電源	+3.3V	3.3V 供電(正)

我們請讀者鍵入 Ｓ ｋ ｅ ｔ ｃ ｈ Ｉ Ｄ Ｅ 軟 體 (軟 體 下 載 請 到 ： https://www.arduino.cc/en/Main/Software)，編譯完成後上傳到開發版進行測試。

表 60 Nokia 5110 LCD 模組測試程式一

Nokia 5110 LCD 模組測試程式一(NOKIA01)

```
#define PIN_SCE     7
#define PIN_RESET 6
#define PIN_DC      5
#define PIN_SDIN    4
#define PIN_SCLK    3


#define LCD_C       LOW
#define LCD_D       HIGH


#define LCD_X       84
#define LCD_Y       48

static const byte ASCII[][5] =
{
  {0x00, 0x00, 0x00, 0x00, 0x00} // 20
 ,{0x00, 0x00, 0x5f, 0x00, 0x00} // 21 !
 ,{0x00, 0x07, 0x00, 0x07, 0x00} // 22 "
 ,{0x14, 0x7f, 0x14, 0x7f, 0x14} // 23 #
 ,{0x24, 0x2a, 0x7f, 0x2a, 0x12} // 24 $
 ,{0x23, 0x13, 0x08, 0x64, 0x62} // 25 %
 ,{0x36, 0x49, 0x55, 0x22, 0x50} // 26 &
 ,{0x00, 0x05, 0x03, 0x00, 0x00} // 27 '
 ,{0x00, 0x1c, 0x22, 0x41, 0x00} // 28 (
 ,{0x00, 0x41, 0x22, 0x1c, 0x00} // 29 )
 ,{0x14, 0x08, 0x3e, 0x08, 0x14} // 2a *
 ,{0x08, 0x08, 0x3e, 0x08, 0x08} // 2b +
 ,{0x00, 0x50, 0x30, 0x00, 0x00} // 2c ,
 ,{0x08, 0x08, 0x08, 0x08, 0x08} // 2d -
 ,{0x00, 0x60, 0x60, 0x00, 0x00} // 2e .
 ,{0x20, 0x10, 0x08, 0x04, 0x02} // 2f /
 ,{0x3e, 0x51, 0x49, 0x45, 0x3e} // 30 0
 ,{0x00, 0x42, 0x7f, 0x40, 0x00} // 31 1
 ,{0x42, 0x61, 0x51, 0x49, 0x46} // 32 2
 ,{0x21, 0x41, 0x45, 0x4b, 0x31} // 33 3
 ,{0x18, 0x14, 0x12, 0x7f, 0x10} // 34 4
 ,{0x27, 0x45, 0x45, 0x45, 0x39} // 35 5
```

```
,{0x3c, 0x4a, 0x49, 0x49, 0x30} // 36 6
,{0x01, 0x71, 0x09, 0x05, 0x03} // 37 7
,{0x36, 0x49, 0x49, 0x49, 0x36} // 38 8
,{0x06, 0x49, 0x49, 0x29, 0x1e} // 39 9
,{0x00, 0x36, 0x36, 0x00, 0x00} // 3a :
,{0x00, 0x56, 0x36, 0x00, 0x00} // 3b ;
,{0x08, 0x14, 0x22, 0x41, 0x00} // 3c <
,{0x14, 0x14, 0x14, 0x14, 0x14} // 3d =
,{0x00, 0x41, 0x22, 0x14, 0x08} // 3e >
,{0x02, 0x01, 0x51, 0x09, 0x06} // 3f ?
,{0x32, 0x49, 0x79, 0x41, 0x3e} // 40 @
,{0x7e, 0x11, 0x11, 0x11, 0x7e} // 41 A
,{0x7f, 0x49, 0x49, 0x49, 0x36} // 42 B
,{0x3e, 0x41, 0x41, 0x41, 0x22} // 43 C
,{0x7f, 0x41, 0x41, 0x22, 0x1c} // 44 D
,{0x7f, 0x49, 0x49, 0x49, 0x41} // 45 E
,{0x7f, 0x09, 0x09, 0x09, 0x01} // 46 F
,{0x3e, 0x41, 0x49, 0x49, 0x7a} // 47 G
,{0x7f, 0x08, 0x08, 0x08, 0x7f} // 48 H
,{0x00, 0x41, 0x7f, 0x41, 0x00} // 49 I
,{0x20, 0x40, 0x41, 0x3f, 0x01} // 4a J
,{0x7f, 0x08, 0x14, 0x22, 0x41} // 4b K
,{0x7f, 0x40, 0x40, 0x40, 0x40} // 4c L
,{0x7f, 0x02, 0x0c, 0x02, 0x7f} // 4d M
,{0x7f, 0x04, 0x08, 0x10, 0x7f} // 4e N
,{0x3e, 0x41, 0x41, 0x41, 0x3e} // 4f O
,{0x7f, 0x09, 0x09, 0x09, 0x06} // 50 P
,{0x3e, 0x41, 0x51, 0x21, 0x5e} // 51 Q
,{0x7f, 0x09, 0x19, 0x29, 0x46} // 52 R
,{0x46, 0x49, 0x49, 0x49, 0x31} // 53 S
,{0x01, 0x01, 0x7f, 0x01, 0x01} // 54 T
,{0x3f, 0x40, 0x40, 0x40, 0x3f} // 55 U
,{0x1f, 0x20, 0x40, 0x20, 0x1f} // 56 V
,{0x3f, 0x40, 0x38, 0x40, 0x3f} // 57 W
,{0x63, 0x14, 0x08, 0x14, 0x63} // 58 X
,{0x07, 0x08, 0x70, 0x08, 0x07} // 59 Y
,{0x61, 0x51, 0x49, 0x45, 0x43} // 5a Z
,{0x00, 0x7f, 0x41, 0x41, 0x00} // 5b [
,{0x02, 0x04, 0x08, 0x10, 0x20} // 5c ¥
```

```
,{0x00, 0x41, 0x41, 0x7f, 0x00} // 5d ]
,{0x04, 0x02, 0x01, 0x02, 0x04} // 5e ^
,{0x40, 0x40, 0x40, 0x40, 0x40} // 5f _
,{0x00, 0x01, 0x02, 0x04, 0x00} // 60 `
,{0x20, 0x54, 0x54, 0x54, 0x78} // 61 a
,{0x7f, 0x48, 0x44, 0x44, 0x38} // 62 b
,{0x38, 0x44, 0x44, 0x44, 0x20} // 63 c
,{0x38, 0x44, 0x44, 0x48, 0x7f} // 64 d
,{0x38, 0x54, 0x54, 0x54, 0x18} // 65 e
,{0x08, 0x7e, 0x09, 0x01, 0x02} // 66 f
,{0x0c, 0x52, 0x52, 0x52, 0x3e} // 67 g
,{0x7f, 0x08, 0x04, 0x04, 0x78} // 68 h
,{0x00, 0x44, 0x7d, 0x40, 0x00} // 69 i
,{0x20, 0x40, 0x44, 0x3d, 0x00} // 6a j
,{0x7f, 0x10, 0x28, 0x44, 0x00} // 6b k
,{0x00, 0x41, 0x7f, 0x40, 0x00} // 6c l
,{0x7c, 0x04, 0x18, 0x04, 0x78} // 6d m
,{0x7c, 0x08, 0x04, 0x04, 0x78} // 6e n
,{0x38, 0x44, 0x44, 0x44, 0x38} // 6f o
,{0x7c, 0x14, 0x14, 0x14, 0x08} // 70 p
,{0x08, 0x14, 0x14, 0x18, 0x7c} // 71 q
,{0x7c, 0x08, 0x04, 0x04, 0x08} // 72 r
,{0x48, 0x54, 0x54, 0x54, 0x20} // 73 s
,{0x04, 0x3f, 0x44, 0x40, 0x20} // 74 t
,{0x3c, 0x40, 0x40, 0x20, 0x7c} // 75 u
,{0x1c, 0x20, 0x40, 0x20, 0x1c} // 76 v
,{0x3c, 0x40, 0x30, 0x40, 0x3c} // 77 w
,{0x44, 0x28, 0x10, 0x28, 0x44} // 78 x
,{0x0c, 0x50, 0x50, 0x50, 0x3c} // 79 y
,{0x44, 0x64, 0x54, 0x4c, 0x44} // 7a z
,{0x00, 0x08, 0x36, 0x41, 0x00} // 7b {
,{0x00, 0x00, 0x7f, 0x00, 0x00} // 7c l
,{0x00, 0x41, 0x36, 0x08, 0x00} // 7d }
,{0x10, 0x08, 0x08, 0x10, 0x08} // 7e ←
,{0x78, 0x46, 0x41, 0x46, 0x78} // 7f →
};

void LcdCharacter(char character)
{
```

```
    LcdWrite(LCD_D, 0x00);
    for (int index = 0; index < 5; index++)
    {
        LcdWrite(LCD_D, ASCII[character - 0x20][index]);
    }
    LcdWrite(LCD_D, 0x00);
}

void LcdClear(void)
{
    for (int index = 0; index < LCD_X * LCD_Y / 8; index++)
    {
        LcdWrite(LCD_D, 0x00);
    }
}

void LcdInitialise(void)
{
    pinMode(PIN_SCE, OUTPUT);
    pinMode(PIN_RESET, OUTPUT);
    pinMode(PIN_DC, OUTPUT);
    pinMode(PIN_SDIN, OUTPUT);
    pinMode(PIN_SCLK, OUTPUT);
    digitalWrite(PIN_RESET, LOW);
    digitalWrite(PIN_RESET, HIGH);
    LcdWrite(LCD_C, 0x21 );   // LCD Extended Commands.
    LcdWrite(LCD_C, 0xB1 );   // Set LCD Vop (Contrast).
    LcdWrite(LCD_C, 0x04 );   // Set Temp coefficent. //0x04
    LcdWrite(LCD_C, 0x14 );   // LCD bias mode 1:48. //0x13
    LcdWrite(LCD_C, 0x20 );   // LCD Basic Commands
    LcdWrite(LCD_C, 0x0C );   // LCD in normal mode.
}

void LcdString(char *characters)
{
    while (*characters)
    {
        LcdCharacter(*characters++);
    }
```

```
    }

    void LcdWrite(byte dc, byte data)
    {
        digitalWrite(PIN_DC, dc);
        digitalWrite(PIN_SCE, LOW);
        shiftOut(PIN_SDIN, PIN_SCLK, MSBFIRST, data);
        digitalWrite(PIN_SCE, HIGH);
    }

    void setup(void)
    {
        LcdInitialise();
        LcdClear();
        LcdString("Hello World!");
    }

    void loop(void)
    {
    }
```

參考資料：http://playground.arduino.cc/Code/PCD8544

程式下載網址：https://github.com/brucetsao/makerdiwo/tree/master/201608

如下圖所示，我們可以看到 Nokia 5110 LCD 模組資訊顯示出來文字(曹永忠,
2016e, 2016f; 曹永忠, 許智誠, et al., 2015f, 2015g, 2015h, 2015i, 2015j, 2015l; 曹永忠,
許碩芳, et al., 2015a, 2015b)。

圖 101 Nokia 5110 LCD 模組測試程式一結果畫面

使用函式庫操控顯示模組顯示圖文字

本段主要介紹讀者，如何使用 Arduino 慣用的函式方式來操控 Nokia 5110 LCD 模組，並顯示圖形、文字、動畫等方式(曹永忠, 2016e, 2016f)。

首先，我們參考 adafruit 官網之學習網站，網址如下：

http://www.adafruit.com/products/338，我們可以到該產品的學習網站：

https://learn.adafruit.com/nokia-5110-3310-monochrome-lcd。

本程式需要用到兩個函式庫(曹永忠, 2016g)，關於 Adafruit_PCD8544.h，請到網址：

https://github.com/brucetsao/LIB_for_MCU/tree/master/Arduino_Lib/libraries/Adafruit-PCD8544-Nokia-5110 或

https://learn.adafruit.com/nokia-5110-3310-monochrome-lcd/graphics-library 或

https://github.com/adafruit/Adafruit-PCD8544-Nokia-5110-LCD-library，進行下載，

關於 Adafruit_GFX.h，請到網址：

https://github.com/brucetsao/LIB_for_MCU/tree/master/Arduino_Lib/libraries/Adafruit_GFX，進行下載。

我們將下表程式，請讀者鍵入Ｓｋｅｔｃｈ　ＩＤＥ軟體(軟體下載請到：

https://www.arduino.cc/en/Main/Software)，編譯完成後上傳到開發版進行測試。

表 61 Nokia 5110 LCD 模組測試程式二

Nokia 5110 LCD 模組測試程式二(NOKIA02)

```
/*********************************************************************
This is an example sketch for our Monochrome Nokia 5110 LCD Displays

    Pick one up today in the adafruit shop!
    ------> http://www.adafruit.com/products/338

These displays use SPI to communicate, 4 or 5 pins are required to
interface

Adafruit invests time and resources providing this open source code,
please support Adafruit and open-source hardware by purchasing
products from Adafruit!

Written by Limor Fried/Ladyada    for Adafruit Industries.
BSD license, check license.txt for more information
All text above, and the splash screen must be included in any redistribution
*********************************************************************/

#include <SPI.h>
#include <Adafruit_GFX.h>
#include <Adafruit_PCD8544.h>
#define PIN_SCE    7
#define PIN_RESET 6
#define PIN_DC     5
#define PIN_SDIN   4
#define PIN_SCLK   3

// Software SPI (slower updates, more flexible pin options):
//=== old version pin out======
// pin 7 - Serial clock out (SCLK)
```

```
// pin 6 - Serial data out (DIN)
// pin 5 - Data/Command select (D/C)
// pin 4 - LCD chip select (CS)
// pin 3 - LCD reset (RST)
Adafruit_PCD8544 display = Adafruit_PCD8544(PIN_SCLK, PIN_SDIN, PIN_DC,
PIN_SCE, PIN_RESET);
// Adafruit_PCD8544 display = Adafruit_PCD8544(SCLK, DIN, D/C, CS/SCE, RST);


#define NUMFLAKES 10
#define XPOS 0
#define YPOS 1
#define DELTAY 2


#define LOGO16_GLCD_HEIGHT 16
#define LOGO16_GLCD_WIDTH   16

static const unsigned char PROGMEM logo16_glcd_bmp[] =
{ B00000000, B11000000,
  B00000001, B11000000,
  B00000001, B11000000,
  B00000011, B11100000,
  B11110011, B11100000,
  B11111110, B11111000,
  B01111110, B11111111,
  B00110011, B10011111,
  B00011111, B11111100,
  B00001101, B01110000,
  B00011011, B10100000,
  B00111111, B11100000,
  B00111111, B11110000,
  B01111100, B11110000,
  B01110000, B01110000,
  B00000000, B00110000 };

void setup()   {
  Serial.begin(9600);
```

```
display.begin();
// init done

// you can change the contrast around to adapt the display
// for the best viewing!
display.setContrast(50);

display.display(); // show splashscreen
delay(2000);
display.clearDisplay();      // clears the screen and buffer

// draw a single pixel
display.drawPixel(10, 10, BLACK);
display.display();
delay(2000);
display.clearDisplay();

// draw many lines
testdrawline();
display.display();
delay(2000);
display.clearDisplay();

// draw rectangles
testdrawrect();
display.display();
delay(2000);
display.clearDisplay();

// draw multiple rectangles
testfillrect();
display.display();
delay(2000);
display.clearDisplay();

// draw mulitple circles
testdrawcircle();
display.display();
delay(2000);
```

```
display.clearDisplay();

// draw a circle, 10 pixel radius
display.fillCircle(display.width()/2, display.height()/2, 10, BLACK);
display.display();
delay(2000);
display.clearDisplay();

testdrawroundrect();
delay(2000);
display.clearDisplay();

testfillroundrect();
delay(2000);
display.clearDisplay();

testdrawtriangle();
delay(2000);
display.clearDisplay();

testfilltriangle();
delay(2000);
display.clearDisplay();

// draw the first ~12 characters in the font
testdrawchar();
display.display();
delay(2000);
display.clearDisplay();

// text display tests
display.setTextSize(1);
display.setTextColor(BLACK);
display.setCursor(0,0);
display.println("Hello, world!");
display.setTextColor(WHITE, BLACK); // 'inverted' text
display.println(3.141592);
display.setTextSize(2);
display.setTextColor(BLACK);
```

```
      display.print("0x"); display.println(0xDEADBEEF, HEX);
      display.display();
      delay(2000);

      // rotation example
      display.clearDisplay();
      display.setRotation(1);    // rotate 90 degrees counter clockwise, can also use values
of 2 and 3 to go further.
      display.setTextSize(1);
      display.setTextColor(BLACK);
      display.setCursor(0,0);
      display.println("Rotation");
      display.setTextSize(2);
      display.println("Example!");
      display.display();
      delay(2000);

      // revert back to no rotation
      display.setRotation(0);

      // miniature bitmap display
      display.clearDisplay();
      display.drawBitmap(30, 16,    logo16_glcd_bmp, 16, 16, 1);
      display.display();

      // invert the display
      display.invertDisplay(true);
      delay(1000);
      display.invertDisplay(false);
      delay(1000);

      // draw a bitmap icon and 'animate' movement
      testdrawbitmap(logo16_glcd_bmp, LOGO16_GLCD_WIDTH,
LOGO16_GLCD_HEIGHT);
    }


    void loop() {
```

```
        }

    void testdrawbitmap(const uint8_t *bitmap, uint8_t w, uint8_t h) {
        uint8_t icons[NUMFLAKES][3];
        randomSeed(666);        // whatever seed

        // initialize
        for (uint8_t f=0; f< NUMFLAKES; f++) {
            icons[f][XPOS] = random(display.width());
            icons[f][YPOS] = 0;
            icons[f][DELTAY] = random(5) + 1;

            Serial.print("x: ");
            Serial.print(icons[f][XPOS], DEC);
            Serial.print(" y: ");
            Serial.print(icons[f][YPOS], DEC);
            Serial.print(" dy: ");
            Serial.println(icons[f][DELTAY], DEC);
        }

        while (1) {
            // draw each icon
            for (uint8_t f=0; f< NUMFLAKES; f++) {
                display.drawBitmap(icons[f][XPOS], icons[f][YPOS], logo16_glcd_bmp, w, h,
BLACK);
            }
            display.display();
            delay(200);

            // then erase it + move it
            for (uint8_t f=0; f< NUMFLAKES; f++) {
                display.drawBitmap(icons[f][XPOS], icons[f][YPOS],   logo16_glcd_bmp, w,
h, WHITE);
                // move it
                icons[f][YPOS] += icons[f][DELTAY];
                // if its gone, reinit
                if (icons[f][YPOS] > display.height()) {
            icons[f][XPOS] = random(display.width());
```

```
      icons[f][YPOS] = 0;
      icons[f][DELTAY] = random(5) + 1;
        }
      }
    }
}


void testdrawchar(void) {
  display.setTextSize(1);
  display.setTextColor(BLACK);
  display.setCursor(0,0);

  for (uint8_t i=0; i < 168; i++) {
    if (i == '\n') continue;
    display.write(i);
    //if ((i > 0) && (i % 14 == 0))
      //display.println();
  }
  display.display();
}

void testdrawcircle(void) {
  for (int16_t i=0; i<display.height(); i+=2) {
    display.drawCircle(display.width()/2, display.height()/2, i, BLACK);
    display.display();
  }
}

void testfillrect(void) {
  uint8_t color = 1;
  for (int16_t i=0; i<display.height()/2; i+=3) {
    // alternate colors
    display.fillRect(i, i, display.width()-i*2, display.height()-i*2, color%2);
    display.display();
    color++;
  }
}
```

```
void testdrawtriangle(void) {
  for (int16_t i=0; i<min(display.width(),display.height())/2; i+=5) {
    display.drawTriangle(display.width()/2, display.height()/2-i,
                         display.width()/2-i, display.height()/2+i,
                         display.width()/2+i, display.height()/2+i, BLACK);
    display.display();
  }
}


void testfilltriangle(void) {
  uint8_t color = BLACK;
  for (int16_t i=min(display.width(),display.height())/2; i>0; i-=5) {
    display.fillTriangle(display.width()/2, display.height()/2-i,
                         display.width()/2-i, display.height()/2+i,
                         display.width()/2+i, display.height()/2+i, color);
    if (color == WHITE) color = BLACK;
    else color = WHITE;
    display.display();
  }
}


void testdrawroundrect(void) {
  for (int16_t i=0; i<display.height()/2-2; i+=2) {
    display.drawRoundRect(i, i, display.width()-2*i, display.height()-2*i,
display.height()/4, BLACK);
    display.display();
  }
}


void testfillroundrect(void) {
  uint8_t color = BLACK;
  for (int16_t i=0; i<display.height()/2-2; i+=2) {
    display.fillRoundRect(i, i, display.width()-2*i, display.height()-2*i,
display.height()/4, color);
    if (color == WHITE) color = BLACK;
    else color = WHITE;
    display.display();
  }
}
```

```
void testdrawrect(void) {
  for (int16_t i=0; i<display.height()/2; i+=2) {
    display.drawRect(i, i, display.width()-2*i, display.height()-2*i, BLACK);
    display.display();
  }
}


void testdrawline() {
  for (int16_t i=0; i<display.width(); i+=4) {
    display.drawLine(0, 0, i, display.height()-1, BLACK);
    display.display();
  }
  for (int16_t i=0; i<display.height(); i+=4) {
    display.drawLine(0, 0, display.width()-1, i, BLACK);
    display.display();
  }
  delay(250);

  display.clearDisplay();
  for (int16_t i=0; i<display.width(); i+=4) {
    display.drawLine(0, display.height()-1, i, 0, BLACK);
    display.display();
  }
  for (int8_t i=display.height()-1; i>=0; i-=4) {
    display.drawLine(0, display.height()-1, display.width()-1, i, BLACK);
    display.display();
  }
  delay(250);

  display.clearDisplay();
  for (int16_t i=display.width()-1; i>=0; i-=4) {
    display.drawLine(display.width()-1, display.height()-1, i, 0, BLACK);
    display.display();
  }
  for (int16_t i=display.height()-1; i>=0; i-=4) {
    display.drawLine(display.width()-1, display.height()-1, 0, i, BLACK);
    display.display();
  }
```

```
    delay(250);

    display.clearDisplay();
    for (int16_t i=0; i<display.height(); i+=4) {
        display.drawLine(display.width()-1, 0, 0, i, BLACK);
        display.display();
    }
    for (int16_t i=0; i<display.width(); i+=4) {
        display.drawLine(display.width()-1, 0, i, display.height()-1, BLACK);
        display.display();
    }
    delay(250);
}
```

程式下載網址：https://github.com/brucetsao/makerdiwo/tree/master/201608

讀者也可以在筆者 YouTube 頻道

(https://www.youtube.com/user/UltimaBruce)中，在網址

https://www.youtube.com/watch?v=NQXnjaP1wc0&feature=youtu.be，看到本次實驗-

Nokia 5110 LCD 模組測試程式二結果畫面。

如下圖所示，我們可以看到 Nokia 5110 LCD 模組資訊顯示出來文字(曹永忠,

2016b, 2016c, 2016e, 2016f; 曹永忠, 許智誠, et al., 2015f, 2015g, 2015h, 2015i, 2015j,

2015l; 曹永忠, 許碩芳, et al., 2015a, 2015b)。

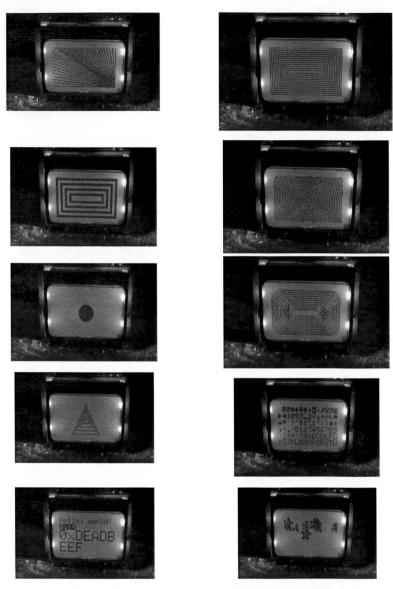

圖 102 Nokia 5110 LCD 模組測試程式二結果畫面

顯示文字

首先，我們要教讀者在 Nokia 5110 LCD 模組顯示文字，首先我們依上面電路
組立將 Nokia 5110 LCD 模組與開發版連接完成，再進行程式寫作。

本程式需要用到兩個函式庫(曹永忠, 2016g)，關於 Adafruit_PCD8544.h，請到網址：

https://github.com/brucetsao/LIB_for_MCU/tree/master/Arduino_Lib/libraries/Adafruit-PCD8544-Nokia-5110 或

https://learn.adafruit.com/nokia-5110-3310-monochrome-lcd/graphics-library 或

https://github.com/adafruit/Adafruit-PCD8544-Nokia-5110-LCD-library，進行下載，關於 Adafruit_GFX.h，請到網址：

https://github.com/brucetsao/LIB_for_MCU/tree/master/Arduino_Lib/libraries/Adafruit_GFX，進行下載。

我們，請讀者鍵入Ｓｋｅｔｃｈ　ＩＤＥ軟體(軟體下載請到：https://www.arduino.cc/en/Main/Software)，編譯完成後上傳到開發版進行測試。

表 62 Nokia 5110 LCD 模組文字測試程式一

Nokia 5110 LCD 模組文字測試程式一(NOKIA_Text_Show)

/***
This is an example sketch for our Monochrome Nokia 5110 LCD Displays
 	
  Pick one up today in the adafruit shop!
  ------> http://www.adafruit.com/products/338
 	
These displays use SPI to communicate, 4 or 5 pins are required to interface
 	
Adafruit invests time and resources providing this open source code,
please support Adafruit and open-source hardware by purchasing
products from Adafruit!
 	
Written by Limor Fried/Ladyada   for Adafruit Industries.
BSD license, check license.txt for more information

All text above, and the splash screen must be included in any redistribution
**/

```
#include <SPI.h>
#include <Adafruit_GFX.h>
#include <Adafruit_PCD8544.h>
#define PIN_SCE    7
#define PIN_RESET 6
#define PIN_DC     5
#define PIN_SDIN   4
#define PIN_SCLK   3


// Software SPI (slower updates, more flexible pin options):
//=== old version pin out======
// pin 7 - Serial clock out (SCLK)
// pin 6 - Serial data out (DIN)
// pin 5 - Data/Command select (D/C)
// pin 4 - LCD chip select (CS)
// pin 3 - LCD reset (RST)
Adafruit_PCD8544 display = Adafruit_PCD8544(PIN_SCLK, PIN_SDIN, PIN_DC,
PIN_SCE, PIN_RESET);
// Adafruit_PCD8544 display = Adafruit_PCD8544(SCLK, DIN, D/C, CS/SCE, RST);


#define NUMFLAKES 10
#define XPOS 0
#define YPOS 1
#define DELTAY 2

void setup() {
  // put your setup code here, to run once:
Serial.begin(9600);

  display.begin();          //init Nokia 5110 display
  // init done

  // you can change the contrast around to adapt the display
  // for the best viewing!
  display.setContrast(50);        //set Contrast
```

```
        display.clearDisplay();            //清除螢幕
        display.display(); // show splashscreen    //


        display.setTextSize(1);      //設定字形大小
        display.setTextColor(BLACK);        //設定字形顏色黑色
        display.setCursor(0,0);      //位置歸零
        display.println("Hello, world!");    //印出字
        display.setTextColor(WHITE, BLACK); // 反白文字
        display.println(3.141592);      //印出字
        display.setTextSize(2);        //設定字形大小
        display.setTextColor(BLACK);          //設定字形顏色黑色
        display.print("0x"); display.println(0xDEADBEEF, HEX);    //印出字
        display.display();        //顯示所有上面內容，必要在所有秀字命令後，一定要
的
        delay(2000);



    }

    void loop() {
        // put your main code here, to run repeatedly:

    }
```

程式下載網址：https://github.com/brucetsao/makerdiwo/tree/master/201608

　　如下圖所示，我們可以看到 Nokia 5110 LCD 模組資訊顯示出來文字(曹永忠,

2016b, 2016c, 2016d, 2016e; 曹永忠, 許智誠, et al., 2015f, 2015g, 2015h, 2015i,

2015j, 2015l; 曹永忠, 許碩芳, et al., 2015a, 2015b)。

圖 103 Nokia 5110 LCD 模組文字測試程式一結果畫面

顯示點幾何圖形

首先,我們要教讀者在 Nokia 5110 LCD 模組畫出許多點,首先我們依上面電路組立將 Nokia 5110 LCD 模組與開發版連接完成,再進行程式寫作。

本程式需要用到兩個函式庫(曹永忠, 2016g),關於 Adafruit_PCD8544.h,請到網址: https://github.com/brucetsao/LIB_for_MCU/tree/master/Arduino_Lib/libraries/Adafruit-PCD8544-Nokia-5110 或 https://learn.adafruit.com/nokia-5110-3310-monochrome-lcd/graphics-library 或 https://github.com/adafruit/Adafruit-PCD8544-Nokia-5110-LCD-library,進行下載,關於 Adafruit_GFX.h,請到網址: https://github.com/brucetsao/LIB_for_MCU/tree/master/Arduino_Lib/libraries/Adafruit_GFX,進行下載。

我 們 請 讀 者 鍵 入 S k e t c h I D E 軟 體 (軟 體 下 載 請 到: https://www.arduino.cc/en/Main/Software),編譯完成後上傳到開發版進行測試。

表 63 Nokia 5110 LCD 模組畫點測試程式一

N Nokia 5110 LCD 模組畫點測試程式一(NOKIA_drawpixel)
/** ** This is an example sketch for our Monochrome Nokia 5110 LCD Displays

Pick one up today in the adafruit shop!
------> http://www.adafruit.com/products/338

These displays use SPI to communicate, 4 or 5 pins are required to
interface

*/

```
#include <SPI.h>
#include <Adafruit_GFX.h>
#include <Adafruit_PCD8544.h>
#define PIN_SCE    7
#define PIN_RESET 6
#define PIN_DC     5
#define PIN_SDIN   4
#define PIN_SCLK   3


// Software SPI (slower updates, more flexible pin options):
//=== old version pin out======
// pin 7 - Serial clock out (SCLK)
// pin 6 - Serial data out (DIN)
// pin 5 - Data/Command select (D/C)
// pin 4 - LCD chip select (CS)
// pin 3 - LCD reset (RST)
Adafruit_PCD8544 display = Adafruit_PCD8544(PIN_SCLK, PIN_SDIN,
PIN_DC, PIN_SCE, PIN_RESET);
// Adafruit_PCD8544 display = Adafruit_PCD8544(SCLK, DIN, D/C, CS/SCE,
RST);


#define NUMFLAKES 10
```

```
#define XPOS 0
#define YPOS 1
#define DELTAY 2

void setup() {
  // put your setup code here, to run once:
Serial.begin(9600);

  display.begin();          //init Nokia 5110 display
  // init done

  // you can change the contrast around to adapt the display
  // for the best viewing!
  display.setContrast(50);          //set Contrast
  display.clearDisplay();           //清除螢幕
  display.display(); // show splashscreen    //

// draw a single pixel
  display.drawPixel(10, 10, BLACK);
  display.drawPixel(10, 12, BLACK);
  display.drawPixel(10, 14, BLACK);
  display.drawPixel(10, 16, BLACK);
  display.drawPixel(10, 18, BLACK);
  display.drawPixel(10, 20, BLACK);
  display.drawPixel(14, 10, BLACK);
  display.drawPixel(14, 12, BLACK);
  display.drawPixel(14, 14, BLACK);
  display.drawPixel(14, 16, BLACK);
  display.drawPixel(14, 18, BLACK);
  display.drawPixel(14, 20, BLACK);
  display.drawPixel(18, 10, BLACK);
  display.drawPixel(18, 12, BLACK);
  display.drawPixel(18, 14, BLACK);
  display.drawPixel(18, 16, BLACK);
  display.drawPixel(18, 18, BLACK);
  display.drawPixel(18, 20, BLACK);
  display.drawPixel(22, 10, BLACK);
  display.drawPixel(22, 12, BLACK);
  display.drawPixel(22, 14, BLACK);
```

```
        display.drawPixel(22, 16, BLACK);
        display.drawPixel(22, 18, BLACK);
        display.drawPixel(22, 20, BLACK);
        display.display();        //顯示所有上面內容，必要在所有秀字命令後，一定
要的
        delay(2000);


    }

    void loop() {
        // put your main code here, to run repeatedly:

    }
```

程式下載網址：https://github.com/brucetsao/makerdiwo/tree/master/201608

如下圖所示，我們可以看到 Nokia 5110 LCD 模組畫出許多點(曹永忠, 2016b, 2016c, 2016d, 2016e, 2016f; 曹永忠, 許智誠, et al., 2015f, 2015g, 2015h, 2015i, 2015j, 2015l; 曹永忠, 許碩芳, et al., 2015a, 2015b)。

圖 104 Nokia 5110 LCD 模組畫點測試程式一結果畫面

顯示線幾何圖形

　　首先，我們要教讀者在 Nokia 5110 LCD 模組畫出線，首先我們依上面電路組立將 Nokia 5110 LCD 模組與開發版連接完成，再進行程式寫作(曹永忠, 2016e, 2016f)。

　　本程式需要用到兩個函式庫(曹永忠, 2016g)，關於 Adafruit_PCD8544.h，請到網址：

https://github.com/brucetsao/LIB_for_MCU/tree/master/Arduino_Lib/libraries/Adafruit-PCD8544-Nokia-5110

或 https://learn.adafruit.com/nokia-5110-3310-monochrome-lcd/graphics-library 或

https://github.com/adafruit/Adafruit-PCD8544-Nokia-5110-LCD-library，進行下載，關於

Adafruit_GFX.h，請到網址：

https://github.com/brucetsao/LIB_for_MCU/tree/master/Arduino_Lib/libraries/Adafruit_GFX

，進行下載。

　　我們請讀者鍵入 Ｓｋｅｔｃｈ　ＩＤＥ軟體 (軟體下載請到：https://www.arduino.cc/en/Main/Software)，編譯完成後上傳到開發版進行測試。

表 64 Nokia 5110 LCD 模組畫線測試程式一

Nokia 5110 LCD 模組畫線測試程式一(NOKIA_Text_Show)
/*** ** This is an example sketch for our Monochrome Nokia 5110 LCD Displays 　Pick one up today in the adafruit shop! 　------> http://www.adafruit.com/products/338 These displays use SPI to communicate, 4 or 5 pins are required to interface Adafruit invests time and resources providing this open source code, please support Adafruit and open-source hardware by purchasing

products from Adafruit!

Written by Limor Fried/Ladyada   for Adafruit Industries.
BSD license, check license.txt for more information
All text above, and the splash screen must be included in any redistribution
****
*/

```
#include <SPI.h>
#include <Adafruit_GFX.h>
#include <Adafruit_PCD8544.h>
#define PIN_SCE    7
#define PIN_RESET 6
#define PIN_DC    5
#define PIN_SDIN  4
#define PIN_SCLK  3


// Software SPI (slower updates, more flexible pin options):
//=== old version pin out======
// pin 7 - Serial clock out (SCLK)
// pin 6 - Serial data out (DIN)
// pin 5 - Data/Command select (D/C)
// pin 4 - LCD chip select (CS)
// pin 3 - LCD reset (RST)
Adafruit_PCD8544 display = Adafruit_PCD8544(PIN_SCLK, PIN_SDIN,
PIN_DC, PIN_SCE, PIN_RESET);
// Adafruit_PCD8544 display = Adafruit_PCD8544(SCLK, DIN, D/C, CS/SCE,
RST);


#define NUMFLAKES 10
#define XPOS 0
#define YPOS 1
#define DELTAY 2

void setup() {
    // put your setup code here, to run once:
Serial.begin(9600);
```

```
        display.begin();           //init Nokia 5110 display
        // init done


        // you can change the contrast around to adapt the display
        // for the best viewing!
        display.setContrast(50);           //set Contrast
        display.clearDisplay();            //清除螢幕
        display.display(); // show splashscreen    //

    // draw a single line
    for (int16_t i=0; i<display.width(); i+=4) {
        display.drawLine(0, 0, i, display.height()-1, BLACK);
        display.display();
    }


        for (int16_t i=0; i<display.height(); i+=4) {
        display.drawLine(0, 0, display.width()-1, i, BLACK);
        display.display();
    }


        display.display();        //顯示所有上面內容，必要在所有秀字命令後，一定
要的
        delay(2000);



    }


    void loop() {
        // put your main code here, to run repeatedly:


    }
```

程式下載網址：https://github.com/brucetsao/makerdiwo/tree/master/201608

如下圖所示，我們可以看到 Nokia 5110 LCD 模組資畫出許多線(曹永忠, 2016b,

2016c, 2016d, 2016e, 2016f; 曹永忠, 許智誠, et al., 2015f, 2015g, 2015h, 2015i, 2015j,

2015l; 曹永忠, 許碩芳, et al., 2015a, 2015b)。

圖 105 Nokia 5110 LCD 模組畫線測試程式一結果畫面

顯示矩形幾何形狀

　　首先，我們要教讀者在 Nokia 5110 LCD 模組畫出幾何圖形：矩形，首先我們

依上面電路組立將 Nokia 5110 LCD 模組與開發版連接完成，再進行程式寫作。

　　本程式需要用到兩個函式庫(曹永忠, 2016g)，關於 Adafruit_PCD8544.h，請到

網址：https://github.com/brucetsao/LIB_for_MCU/tree/master/Arduino_Lib/libraries/Adafruit-PCD8544-Nokia-5110 或

https://learn.adafruit.com/nokia-5110-3310-monochrome-lcd/graphics-library 或

https://github.com/adafruit/Adafruit-PCD8544-Nokia-5110-LCD-library，進行下載，

關於 Adafruit_GFX.h，請到網址：

https://github.com/brucetsao/LIB_for_MCU/tree/master/Arduino_Lib/libraries/Adafruit

_GFX，進行下載。

　　我們請讀者鍵入Ｓｋｅｔｃｈ　ＩＤＥ軟體（軟體下載請到：

https://www.arduino.cc/en/Main/Software)，編譯完成後上傳到開發版進行測試。

表 65 Nokia 5110 LCD 模組矩形幾何形狀測試程式一

Nokia 5110 LCD 模組矩形幾何形狀測試程式一(NOKIA_drawGEO)
/***

```
**
    This is an example sketch for our Monochrome Nokia 5110 LCD Displays

      Pick one up today in the adafruit shop!
      ------> http://www.adafruit.com/products/338

    These displays use SPI to communicate, 4 or 5 pins are required to
    interface

    Adafruit invests time and resources providing this open source code,
    please support Adafruit and open-source hardware by purchasing
    products from Adafruit!

    Written by Limor Fried/Ladyada    for Adafruit Industries.
    BSD license, check license.txt for more information
    All text above, and the splash screen must be included in any redistribution
    ****************************************************************
*/

    #include <SPI.h>
    #include <Adafruit_GFX.h>
    #include <Adafruit_PCD8544.h>
    #define PIN_SCE    7
    #define PIN_RESET 6
    #define PIN_DC    5
    #define PIN_SDIN   4
    #define PIN_SCLK   3

    // Software SPI (slower updates, more flexible pin options):
    //=== old version pin out======
    // pin 7 - Serial clock out (SCLK)
    // pin 6 - Serial data out (DIN)
    // pin 5 - Data/Command select (D/C)
    // pin 4 - LCD chip select (CS)
    // pin 3 - LCD reset (RST)
    Adafruit_PCD8544 display = Adafruit_PCD8544(PIN_SCLK, PIN_SDIN,
PIN_DC, PIN_SCE, PIN_RESET);
    // Adafruit_PCD8544 display = Adafruit_PCD8544(SCLK, DIN, D/C, CS/SCE,
RST);
```

```
#define NUMFLAKES 10
#define XPOS 0
#define YPOS 1
#define DELTAY 2

void setup() {
    // put your setup code here, to run once:
Serial.begin(9600);

    display.begin();          //init Nokia 5110 display
    // init done

    // you can change the contrast around to adapt thc display
    // for the best viewing!
    display.setContrast(50);          //set Contrast
    display.clearDisplay();           //清除螢幕
    display.display(); // show splashscreen    //

// draw a  多個矩形
    for (int16_t i=0; i<display.height()/2; i+=2) {
        display.drawRect(i, i, display.width()-2*i, display.height()-2*i, BLACK);
        display.display();
    }

    display.display();        //顯示所有上面內容，必要在所有秀字命令後，一定
要的
    delay(2000);



}

void loop() {
    // put your main code here, to run repeatedly:

}
```

如下圖所示，我們可以看到 Nokia 5110 LCD 模組畫出幾何圖形：矩形(曹永忠,
2016b, 2016c, 2016d, 2016e, 2016f; 曹永忠, 許智誠, et al., 2015f, 2015g, 2015h, 2015i,
2015j, 2015l; 曹永忠, 許碩芳, et al., 2015a, 2015b)。

圖 106 Nokia 5110 LCD 模組矩形幾何形狀測試程式一結果畫面

顯示圓形幾何形狀

首先，我們要教讀者在 Nokia 5110 LCD 模組畫出幾何圖形：圓形，首先我們
依上面電路組立將 Nokia 5110 LCD 模組與開發版連接完成，再進行程式寫作。

本程式需要用到兩個函式庫(曹永忠, 2016g)，關於 Adafruit_PCD8544.h，請到網
址：https://github.com/brucetsao/LIB_for_MCU/tree/master/Arduino_Lib/libraries/Adafruit-PCD8544-Nokia-5110 或
https://learn.adafruit.com/nokia-5110-3310-monochrome-lcd/graphics-library 或
https://github.com/adafruit/Adafruit-PCD8544-Nokia-5110-LCD-library，進行下載，關於
Adafruit_GFX.h，請到網址：
https://github.com/brucetsao/LIB_for_MCU/tree/master/Arduino_Lib/libraries/Adafruit_GFX
，進行下載。

我們請讀者鍵入 Ｓ ｋ ｅ ｔ ｃ ｈ　Ｉ Ｄ Ｅ 軟體 （軟體下載請到：
https://www.arduino.cc/en/Main/Software)，編譯完成後上傳到開發版進行測試。

表 66 Nokia 5110 LCD 模組圓形幾何形狀測試程式一

Nokia 5110 LCD 模組圓形幾何形狀測試程式一(NOKIA_drawGEO1)

```
/*****************************************************************
This is an example sketch for our Monochrome Nokia 5110 LCD Displays

  Pick one up today in the adafruit shop!
  ------> http://www.adafruit.com/products/338

These displays use SPI to communicate, 4 or 5 pins are required to
interface

Adafruit invests time and resources providing this open source code,
please support Adafruit and open-source hardware by purchasing
products from Adafruit!

Written by Limor Fried/Ladyada    for Adafruit Industries.
BSD license, check license.txt for more information
All text above, and the splash screen must be included in any redistribution
*****************************************************************/


#include <SPI.h>
#include <Adafruit_GFX.h>
#include <Adafruit_PCD8544.h>
#define PIN_SCE     7
#define PIN_RESET 6
#define PIN_DC      5
#define PIN_SDIN    4
#define PIN_SCLK    3

// Software SPI (slower updates, more flexible pin options):
//=== old version pin out======
// pin 7 - Serial clock out (SCLK)
// pin 6 - Serial data out (DIN)
// pin 5 - Data/Command select (D/C)
// pin 4 - LCD chip select (CS)
// pin 3 - LCD reset (RST)
Adafruit_PCD8544 display = Adafruit_PCD8544(PIN_SCLK, PIN_SDIN, PIN_DC,
PIN_SCE, PIN_RESET);
```

```
// Adafruit_PCD8544 display = Adafruit_PCD8544(SCLK, DIN, D/C, CS/SCE, RST);


#define NUMFLAKES 10
#define XPOS 0
#define YPOS 1
#define DELTAY 2

void setup() {
  // put your setup code here, to run once:
Serial.begin(9600);

  display.begin();          //init Nokia 5110 display
  // init done

  // you can change the contrast around to adapt the display
  // for the best viewing!
  display.setContrast(50);          //set Contrast
  display.clearDisplay();           //清除螢幕
  display.display(); // show splashscreen    //

// draw a  多個矩形
  for (int16_t i=0; i<display.height(); i+=6) {
    display.drawCircle(display.width()/2, display.height()/2, i/2, BLACK);
    display.display();
  }

  display.display();        //顯示所有上面內容，必要在所有秀字命令後，一定要的
  delay(2000);



}


void loop() {
  // put your main code here, to run repeatedly:

}
```

程式下載網址：https://github.com/brucetsao/makerdiwo/tree/master/201608

如下圖所示，我們可以看到 Nokia 5110 LCD 模組畫出幾何圖形：圓形形狀(曹永忠, 2016b, 2016c, 2016d, 2016e, 2016f; 曹永忠, 許智誠, et al., 2015f, 2015g, 2015h, 2015i, 2015j, 2015l; 曹永忠, 許碩芳, et al., 2015a, 2015b)。

圖 107 Nokia 5110 LCD 模組圓形幾何形狀測試程式一結果畫面

顯示三角形幾何形狀

首先，我們要教讀者在 Nokia 5110 LCD 模組畫出幾何圖形：三角形幾何形狀，首先我們依上面電路組立將 Nokia 5110 LCD 模組與開發版連接完成，再進行程式寫作。

本程式需要用到兩個函式庫(曹永忠, 2016g)，關於 Adafruit_PCD8544.h，請到網址：

https://github.com/brucetsao/LIB_for_MCU/tree/master/Arduino_Lib/libraries/Adafruit-PCD8544-Nokia-5110 或

https://learn.adafruit.com/nokia-5110-3310-monochrome-lcd/graphics-library 或

https://github.com/adafruit/Adafruit-PCD8544-Nokia-5110-LCD-library，進行下載，關於 Adafruit_GFX.h，請到網址：

https://github.com/brucetsao/LIB_for_MCU/tree/master/Arduino_Lib/libraries/Adafruit_GFX

，進行下載。

我們請讀者鍵入Ｓｋｅｔｃｈ　ＩＤＥ軟體（軟體下載請到：
https://www.arduino.cc/en/Main/Software)，編譯完成後上傳到開發版進行測試。

表 67 Nokia 5110 LCD 模組畫三角形幾何形狀測試程式一

Nokia 5110 LCD 模組畫三角形幾何形狀測試程式一(NOKIA_drawGEO2)

```
/********************************************************************
This is an example sketch for our Monochrome Nokia 5110 LCD Displays

    Pick one up today in the adafruit shop!
    ------> http://www.adafruit.com/products/338

These displays use SPI to communicate, 4 or 5 pins are required to
interface

Adafruit invests time and resources providing this open source code,
please support Adafruit and open-source hardware by purchasing
products from Adafruit!

Written by Limor Fried/Ladyada    for Adafruit Industries.
BSD license, check license.txt for more information
All text above, and the splash screen must be included in any redistribution
********************************************************************/

#include <SPI.h>
#include <Adafruit_GFX.h>
#include <Adafruit_PCD8544.h>
#define PIN_SCE     7
#define PIN_RESET 6
#define PIN_DC      5
#define PIN_SDIN    4
#define PIN_SCLK    3

// Software SPI (slower updates, more flexible pin options):
//=== old version pin out======
// pin 7 - Serial clock out (SCLK)
```

```
// pin 6 - Serial data out (DIN)
// pin 5 - Data/Command select (D/C)
// pin 4 - LCD chip select (CS)
// pin 3 - LCD reset (RST)
Adafruit_PCD8544 display = Adafruit_PCD8544(PIN_SCLK, PIN_SDIN, PIN_DC,
PIN_SCE, PIN_RESET);
// Adafruit_PCD8544 display = Adafruit_PCD8544(SCLK, DIN, D/C, CS/SCE, RST);


#define NUMFLAKES 10
#define XPOS 0
#define YPOS 1
#define DELTAY 2

void setup() {
    // put your setup code here, to run once:
Serial.begin(9600);

    display.begin();          //init Nokia 5110 display
    // init done

    // you can change the contrast around to adapt the display
    // for the best viewing!
    display.setContrast(50);          //set Contrast
    display.clearDisplay();          //清除螢幕
    display.display(); // show splashscreen    //

    // draw a 多個矩形

    for (int16_t i=0; i<display.height()/2; i+=4)
    {
    //      display.drawTriangle(X0,Y0,X1,Y1,X2,Y2,顏色);
        display.drawTriangle(display.width()/2, i, 0+i*2, display.height()-2-i,
display.width()-i*2, display.height()-2-i,    BLACK);
        display.display();
        // delay(2000) ;
    }

    display.display();      //顯示所有上面內容，必要在所有秀字命令後，一定要的
```

```
    delay(2000);


}

void loop() {
    // put your main code here, to run repeatedly:


}
```

如下圖所示，我們可以看到 Nokia 5110 LCD 模組畫出幾何圖形：三角形幾何

形狀(曹永忠, 2016b, 2016c, 2016d, 2016e, 2016f; 曹永忠, 許智誠, et al., 2015f, 2015g,

2015h, 2015i, 2015j, 2015l; 曹永忠, 許碩芳, et al., 2015a, 2015b)。

圖 108 Nokia 5110 LCD 模組畫三角形幾何形狀測試程式一結果畫面

Nokia 5110 LCD 基礎函數用法

為了更能了解 Nokia 5110 LCD 的用法，本節詳細介紹了 Nokia 5110 LCD 函

式主要的用法：

產生 Nokia Lcd 物件

```
                              LCD5110(SCK, MOSI, DC, RST, CS);
The main class constructor.

Parameters:       SCK:      Pin for Clock signal
                  MOSI:     Pin for Data transfer
                  DC:       Pin for Register Select (Data/Command)
                  RST:      Pin for Reset
                  CS:       Pin for Chip Select
Usage:            LCD5110 myGLCD(8, 9, 10, 11, 12); // Start an instance of the LCD5110 class
```

資料來源：Rinky-Dink Electronics，http://www.rinkydinkelectronics.com/library.php?id=44

初始化 *Nokia Lcd* 物件

```
                              InitLCD([contrast]);
Initialize the LCD.

Parameters:       contrast:  <optional>
                             Specify a value to use for contrast (0-127)
                             Default is 70
Usage:            myGLCD.initLCD(); // Initialize the display
Notes:            This will reset and clear the display.
```

資料來源：Rinky-Dink Electronics，http://www.rinkydinkelectronics.com/library.php?id=44

設定 *Nokia Lcd* 畫面對比

```
                              setContrast(contrast);
Set the contrast of the LCD.

Parameters:       contrast:  Specify a value to use for contrast (0-127)
Usage:            myGLCD.setContrast(70); // Sets the contrast to the default value of 70
```

資料來源：Rinky-Dink Electronics，http://www.rinkydinkelectronics.com/library.php?id=44

設定 *Nokia Lcd* 物件進入睡眠狀態

```
                              enableSleep();
Put the display in Sleep Mode.

Parameters:       None
Usage:            myGLCD.enableSleep(); // Put the display into Sleep Mode
Notes:            Entering Sleep Mode will not turn off the backlight as this is a hardware function.
```

資料來源：Rinky-Dink Electronics，http://www.rinkydinkelectronics.com/library.php?id=44

解除 *Nokia Lcd* 物件睡眠狀態，開始工作

```
                              disableSleep();
Re-enable the display after it has been put in Sleep Mode.

Parameters:       None
Usage:            myGLCD.disableSleep(); // Wake the display after putting it into Sleep Mode
Notes:            The display will automatically be cleared when Sleep Mode is disabled.
                  Exiting Sleep Mode will not turn on the backlight as this is a hardware function.
```

清除畫面

clrScr();
Clear the screen.

Parameters:	None
Usage:	myGLCD.clrScr(); // Clear the screen

清除一部分畫面或全部畫面

clrRow(row[, start_x[, end_x]]);
Clear a part of, or a whole row.

Parameters:	row:	8 pixel high row to clear (0-5)
	start_x:	<optional>
		x-coordinate to start the clearing on (default = 0)
	end_x:	<optional>
		x-coordinate to end the clearing on (default = 83)
Usage:	myGLCD.clrRow(5, 42); // Clear the right half of the lower row	

反向畫面

invert(mode);
Set inversion of the display on or off.

Parameters:	mode: true - Invert the display
	false - Normal display
Usage:	myGLCD.invert(true); // Set display inversion on

在指定位置上印出字串

print(st, x, y);
Print a string at the specified coordinates.
You can use the literals LEFT, CENTER and RIGHT as the x-coordinate to align the string on the screen.

Parameters:	st: the string to print
	x: x-coordinate of the upper, left corner of the first character
	y: y-coordinate of the upper, left corner of the first character
Usage:	myGLCD.print("Hello World",CENTER,0); // Print "Hello World" centered at the top of the screen
Notes:	The y-coordinate will be adjusted to be aligned with an 8 pixel high display row.
	In effect only 0, 8, 16, 24, 32 and 40 can be used as y-coordinates.
	The string can be either a char array or a String object

在指定位置上印出整數數字

printNumI(num, x, y[, length[, filler]]);
Print an integer number at the specified coordinates. You can use the literals LEFT, CENTER and RIGHT as the x-coordinate to align the string on the screen.

Parameters:	num: the value to print (-2,147,483,648 to 2,147,483,647) *INTEGERS ONLY*
	x: x-coordinate of the upper, left corner of the first digit/sign
	y: y-coordinate of the upper, left corner of the first digit/sign
	length: **<optional>**
	        minimum number of digits/characters (including sign) to display
	filler: **<optional>**
	        filler character to use to get the minimum length. The character will be inserted in front
	        of the number, but after the sign. Default is ' ' (space).

Usage:	myGLCD.print(num,CENTER,0); // Print the value of "num" centered at the top of the screen
Notes:	The y-coordinate will be adjusted to be aligned with an 8 pixel high display row. In effect only 0, 8, 16, 24, 32 and 40 can be used as y-coordinates.

資料來源：Rinky-Dink Electronics，http://www.rinkydinkelectronics.com/library.php?id=44

在指定位置上印出實數數字

printNumF(num, dec, x, y[, divider[, length[, filler]]]);
Print a floating-point number at the specified coordinates. You can use the literals LEFT, CENTER and RIGHT as the x-coordinate to align the string on the screen. **WARNING**: Floating point numbers are not exact, and may yield strange results when compared. Use at your own discretion.

Parameters:	num: the value to print (*See note*)
	dec: digits in the fractional part (1-5) *0 is not supported. Use printNumI() instead.*
	x: x-coordinate of the upper, left corner of the first digit/sign
	y: y-coordinate of the upper, left corner of the first digit/sign
	divider: **<Optional>**
	        Single character to use as decimal point. Default is '.'
	length: **<optional>**
	        minimum number of digits/characters (including sign) to display
	filler: **<optional>**
	        filler character to use to get the minimum length. The character will be inserted in front
	        of the number, but after the sign. Default is ' ' (space).

Usage:	myGLCD.print(num, 3, CENTER,0); // Print the value of "num" with 3 fractional digits top centered
Notes:	Supported range depends on the number of fractional digits used. Approx range is +/- 2*(10^(9-dec)) The y-coordinate will be adjusted to be aligned with an 8 pixel high display row. In effect only 0, 8, 16, 24, 32 and 40 can be used as y-coordinates.

資料來源：Rinky-Dink Electronics，http://www.rinkydinkelectronics.com/library.php?id=44

設定反向文字輸出

invertText(mode);
Select if text printed with print(), printNumI() and printNumF() should be inverted.

Parameters:	mode: true - Invert the text false - Normal text
Usage:	myGLCD.invertText(true); // Turn on inverted printing
Notes:	SetFont() will turn off inverted printing

資料來源：Rinky-Dink Electronics，http://www.rinkydinkelectronics.com/library.php?id=44

設定字形

setFont(fontname);
Select font to use with print(), printNumI() and printNumF().

Parameters:	fontname: Name of the array containing the font you wish to use
Usage:	myGLCD.setFont(SmallFont); // Select the font called SmallFont
Notes:	You must declare the font-array as an external or include it in your sketch.

資料來源：Rinky-Dink Electronics，http://www.rinkydinkelectronics.com/library.php?id=44

在指定位置上畫出圖片(黑白)

drawBitmap (x, y, data, sx, sy);
Draw a bitmap on the screen.

Parameters:	x:	x-coordinate of the upper, left corner of the bitmap
	y:	y-coordinate of the upper, left corner of the bitmap
	data:	array containing the bitmap-data
	sx:	width of the bitmap in pixels
	sy:	height of the bitmap in pixels
Usage:	myGLCD.drawBitmap(0, 0, bitmap, 32, 32); // Draw a 32x32 pixel bitmap in the upper left corner
Notes:	You can use the online-tool "ImageConverter Mono" to convert pictures into compatible arrays. The online-tool can be found on my website. Requires that you #include <avr/pgmspace.h> when using an Arduino other than Arduino Due.

資料來源：Rinky-Dink Electronics，http://www.rinkydinkelectronics.com/library.php?id=44

Nokia 5110 LCD 圖形函數用法

為了進一步了解 Nokia 5110 LCD 的用法，本節詳細介紹了 Nokia 5110 LCD 圖形函式主要的用法：

產生 Nokia Lcd 物件

LCD5110(SCK, MOSI, DC, RST, CS);
The main class constructor.

Parameters:	SCK:	Pin for Clock signal
	MOSI:	Pin for Data transfer
	DC:	Pin for Register Select (Data/Command)
	RST:	Pin for Reset
	CS:	Pin for Chip Select
Usage:	LCD5110 myGLCD(8, 9, 10, 11, 12); // Start an instance of the LCD5110 class

資料來源：Rinky-Dink Electronics，http://www.rinkydinkelectronics.com/library.php?id=44

初始化 Nokia Lcd 物件

InitLCD([contrast]);
Initialize the LCD.

Parameters: contrast: \<optional>
                      Specify a value to use for contrast (0-127)
                      Default is 70

Usage: myGLCD.initLCD(); // Initialize the display
Notes: This will reset and clear the display.

資料來源：Rinky-Dink Electronics，http://www.rinkydinkelectronics.com/library.php?id=44

設定 *Nokia Lcd 畫面對比*

setContrast(contrast);
Set the contrast of the LCD.
Parameters: contrast: Specify a value to use for contrast (0-127)
Usage: myGLCD.setContrast(70); // Sets the contrast to the default value of 70

資料來源：Rinky-Dink Electronics，http://www.rinkydinkelectronics.com/library.php?id=44

設定 *Nokia Lcd 物件進入睡眠狀態*

enableSleep();
Put the display in Sleep Mode.
Parameters: None
Usage: myGLCD.enableSleep(); // Put the display into Sleep Mode
Notes: update() will not work while the display is in Sleep Mode.   Entering Sleep Mode will not turn off the backlight as this is a hardware function.

資料來源：Rinky-Dink Electronics，http://www.rinkydinkelectronics.com/library.php?id=44

解除 *Nokia Lcd 物件睡眠狀態，開始工作*

disableSleep();
Re-enable the display after it has been put in Sleep Mode.
Parameters: None
Usage: myGLCD.disableSleep(); // Wake the display after putting it into Sleep Mode

Notes: The display will automatically be updated with the contents of the buffer when Sleep Mode is
                      disabled.
                      Exiting Sleep Mode will not turn on the backlight as this is a hardware function.

資料來源：Rinky-Dink Electronics，http://www.rinkydinkelectronics.com/library.php?id=44

將先前繪圖指令結果，顯示在畫面上

update();
Copy the screen buffer to the screen.
This is the only command, except invert(), that will make anything happen on the physical screen. All other commands only modify the screen buffer.
Parameters: None
Usage: myGLCD.update(); // Copy the screen buffer to the screen
Notes: Remember to call update() after you have updated the screen buffer. Calling update() while the display is in Sleep Mode will not have any effect.

資料來源：Rinky-Dink Electronics，http://www.rinkydinkelectronics.com/library.php?id=44

清除畫面

clrScr();
Clear the screen buffer.
Parameters: None
Usage: myGLCD.clrScr(); // Clear the screen buffer

資料來源：Rinky-Dink Electronics，http://www.rinkydinkelectronics.com/library.php?id=44

將繪圖緩衝區內容填入畫面上

fillScr();
Fill the screen buffer.
Parameters: None
Usage: myGLCD.fillScr(); // Fill the screen buffer

資料來源：Rinky-Dink Electronics，http://www.rinkydinkelectronics.com/library.php?id=44

反向畫面

invert(mode);
Set inversion of the display on or off.
Parameters: mode: true - Invert the display false - Normal display
Usage: myGLCD.invert(true); // Set display inversion on

資料來源：Rinky-Dink Electronics，http://www.rinkydinkelectronics.com/library.php?id=44

在指定位置上畫出一點

setPixel(x, y);		
Turn on the specified pixel in the screen buffer.		
Parameters:	x:	x-coordinate of the pixel
	y:	y-coordinate of the pixel
Usage:		myGLCD.setPixel(0, 0); // Turn on the upper left pixel (in the screen buffer)

資料來源：Rinky-Dink Electronics，http://www.rinkydinkelectronics.com/library.php?id=44

在指定位置上清除一點

clrPixel(x, y);		
Turn off the specified pixel in the screen buffer.		
Parameters:	x:	x-coordinate of the pixel
	y:	y-coordinate of the pixel
Usage:		myGLCD.clrPixel(0, 0); // Turn off the upper left pixel (in the screen buffer)

資料來源：Rinky-Dink Electronics，http://www.rinkydinkelectronics.com/library.php?id=44

在指定位置上那點反向

invPixel(x, y);		
Invert the state of the specified pixel in the screen buffer.		
Parameters:	x:	x-coordinate of the pixel
	y:	y-coordinate of the pixel
Usage:		myGLCD.invPixel(0, 0); // Invert the upper left pixel (in the screen buffer)

資料來源：Rinky-Dink Electronics，http://www.rinkydinkelectronics.com/library.php?id=44

在指定位置上印出字串

print(st, x, y);		
Print a string at the specified coordinates in the screen buffer.		
You can use the literals LEFT, CENTER and RIGHT as the x-coordinate to align the string on the screen.		
Parameters:	st:	the string to print
	x:	x-coordinate of the upper, left corner of the first character
	y:	y-coordinate of the upper, left corner of the first character
Usage:		myGLCD.print("Hello World",CENTER,0); // Print "Hello World" centered at the top of the screen (in the screen buffer)
Notes:		The string can be either a char array or a String object

資料來源：Rinky-Dink Electronics，http://www.rinkydinkelectronics.com/library.php?id=44

在指定位置上印出整數數字

printNumI(num, x, y[, length[, filler]]);
Print an integer number at the specified coordinates in the screen buffer. You can use the literals LEFT, CENTER and RIGHT as the x-coordinate to align the string on the screen.

```
Parameters:    num: the value to print (-2,147,483,648 to 2,147,483,647) INTEGERS ONLY
               x:   x-coordinate of the upper, left corner of the first digit/sign
               y:   y-coordinate of the upper, left corner of the first digit/sign
               length: <optional>
                       minimum number of digits/characters (including sign) to display
               filler: <optional>
                       filler character to use to get the minimum length. The character will be inserted in front
                       of the number, but after the sign. Default is ' ' (space).
Usage:         myGLCD.print(num,CENTER,0); // Print the value of "num" centered at the top of the screen (in the
               screen buffer)
```

資料來源：Rinky-Dink Electronics，http://www.rinkydinkelectronics.com/library.php?id=44

在指定位置上印出實數數字

printNumF(num, dec, x, y[, divider[, length[, filler]]]);
Print a floating-point number at the specified coordinates in the screen buffer. You can use the literals LEFT, CENTER and RIGHT as the x-coordinate to align the string on the screen. **WARNING:** Floating point numbers are not exact, and may yield strange results when compared. Use at your own discretion.

```
Parameters:    num: the value to print (See note)
               dec: digits in the fractional part (1-5) 0 is not supported. Use printNumI() instead.
               x:   x-coordinate of the upper, left corner of the first digit/sign
               y:   y-coordinate of the upper, left corner of the first digit/sign
               divider: <Optional>
                        Single character to use as decimal point. Default is '.'
               length:  <optional>
                        minimum number of digits/characters (including sign) to display
               filler:  <optional>
                        filler character to use to get the minimum length. The character will be inserted in front
                        of the number, but after the sign. Default is ' ' (space).
Usage:         myGLCD.print(num, 3, CENTER,0); // Print the value of "num" with 3 fractional digits top centered
               (in the screen buffer)
Notes:         Supported range depends on the number of fractional digits used.
               Approx range is +/- 2*(10^(9-dec))
```

資料來源：Rinky-Dink Electronics，http://www.rinkydinkelectronics.com/library.php?id=44

設定文字輸出反向顯示

invertText(mode);
Select if text printed with print(), printNumI() and printNumF() should be inverted.

```
Parameters:    mode: true  - Invert the text
                     false - Normal text
Usage:         myGLCD.invertText(true); // Turn on inverted printing
Notes:         SetFont() will turn off inverted printing
```

資料來源：Rinky-Dink Electronics，http://www.rinkydinkelectronics.com/library.php?id=44

設定字形

setFont(fontname);
Select font to use with print(), printNumI() and printNumF().

```
Parameters:    fontname: Name of the array containing the font you wish to use
Usage:         myGLCD.setFont(SmallFont); // Select the font called SmallFont
Notes:         You must declare the font-array as an external or include it in your sketch.
```

在指定兩點(X1,Y1)-(X2,Y2) 畫出一條線

```
                            drawLine(x1, y1, x2, y2);
Draw a line between two points in the screen buffer.

Parameters:    x1: x-coordinate of the start-point
               y1: y-coordinate of the start-point
               x2: x-coordinate of the end-point
               y2: y-coordinate of the end-point
Usage:         myGLCD.drawLine(0,0,83,47); // Draw a line from the upper left to the lower right corner
```

清除指定兩點(X1,Y1)-(X2,Y2) 的一條線

```
                            clrLine(x1, y1, x2, y2);
Clear a line between two points in the screen buffer.

Parameters:    x1: x-coordinate of the start-point
               y1: y-coordinate of the start-point
               x2: x-coordinate of the end-point
               y2: y-coordinate of the end-point
Usage:         myGLCD.clrLine(0,0,83,47); // Clear a line from the upper left to the lower right corner
```

畫出指定對角兩點(X1,Y1)-(X2,Y2) 的矩形

```
                            drawRect(x1, y1, x2, y2);
Draw a rectangle between two points in the screen buffer.

Parameters:    x1: x-coordinate of the start-corner
               y1: y-coordinate of the start-corner
               x2: x-coordinate of the end-corner
               y2: y-coordinate of the end-corner
Usage:         myGLCD.drawRect(42,24,83,47); // Draw a rectangle in the lower right corner of the screen
```

清除指定對角兩點(X1,Y1)-(X2,Y2) 的矩形

```
                            clrRect(x1, y1, x2, y2);
Clear a rectangle between two points in the screen buffer.

Parameters:    x1: x-coordinate of the start-corner
               y1: y-coordinate of the start-corner
               x2: x-coordinate of the end-corner
               y2: y-coordinate of the end-corner
Usage:         myGLCD.clrRect(42,24,83,47); // Clear a rectangle in the lower right corner of the screen
```

畫出指定對角兩點(X1,Y1)-(X2,Y2) 的圓角矩形

drawRoundRect(x1, y1, x2, y2);
Draw a rectangle with slightly rounded corners between two points in the screen buffer. The minimum size is 5 pixels in both directions. If a smaller size is requested the rectangle will not be drawn.

Parameters: x1: x-coordinate of the start-corner
                 y1: y-coordinate of the start-corner
                 x2: x-coordinate of the end-corner
                 y2: y-coordinate of the end-corner

Usage: myGLCD.drawRoundRect(0,0,41,23); // Draw a rounded rectangle in the upper left corner of the screen

資料來源：Rinky-Dink Electronics，http://www.rinkydinkelectronics.com/library.php?id=44

清除指定對角兩點(X1,Y1)-(X2,Y2) 的圓角矩形

clrRoundRect(x1, y1, x2, y2);
Clear a rectangle with slightly rounded corners between two points in the screen buffer. The minimum size is 5 pixels in both directions. If a smaller size is requested the rectangle will not be drawn/cleared.

Parameters: x1: x-coordinate of the start-corner
                 y1: y-coordinate of the start-corner
                 x2: x-coordinate of the end-corner
                 y2: y-coordinate of the end-corner

Usage: myGLCD.clrRoundRect(0,0,41,23); // Clear a rounded rectangle in the upper left corner of the screen

資料來源：Rinky-Dink Electronics，http://www.rinkydinkelectronics.com/library.php?id=44

畫出圓心(X1,Y1)，半徑 RADIUS 的圓形

drawCircle(x, y, radius);
Draw a circle with a specified radius in the screen buffer.

Parameters: x: x-coordinate of the center of the circle
                 y: y-coordinate of the center of the circle
                 radius: radius of the circle in pixels

Usage: myGLCD.drawCircle(41,23,20); // Draw a circle in the middle of the screen with a radius of 20 pixels

資料來源：Rinky-Dink Electronics，http://www.rinkydinkelectronics.com/library.php?id=44

清除圓心(X1,Y1)，半徑 RADIUS 的圓形

clrCircle(x, y, radius);
Clear a circle with a specified radius in the screen buffer.

Parameters: x: x-coordinate of the center of the circle
                 y: y-coordinate of the center of the circle
                 radius: radius of the circle in pixels

Usage: myGLCD.clrCircle(41,23,20); // Clear a circle in the middle of the screen with a radius of 20 pixels

資料來源：Rinky-Dink Electronics，http://www.rinkydinkelectronics.com/library.php?id=44

在指定位置上畫出圖片(黑白)

資料來源：Rinky-Dink Electronics，http://www.rinkydinkelectronics.com/library.php?id=44

章節小結

本章主要介紹之 Arduino 開發板使用與連接 NOKIA 5110 LCD 顯示模組，透過本章節的解說，相信讀者會對連接、使用 NOKIA 5110 LCD 顯示模組，有更深入的了解與體認。

本書總結

筆者對於 Arduino 相關的書籍，也出版許多書籍，感謝許多有心的讀者提供筆者許多寶貴的意見與建議，筆者群不勝感激，許多讀者希望筆者可以推出更多的入門書籍給更多想要進入『Arduino』、『Maker』這個未來大趨勢，所有才有這個入門系列的產生。

本系列叢書的特色是一步一步教導大家使用更基礎的東西，來累積各位的基礎能力，讓大家能更在 Maker 自造者運動中，可以拔的頭籌，所以本系列是一個永不結束的系列，只要更多的東西被製造出來，相信筆者會更衷心的希望與各位永遠在這條 Maker 路上與大家同行。

作者介紹

曹永忠 (Yung-Chung Tsao) ，目前為自由作家暨專業 Maker，專研於軟體工程、軟體開發與設計、物件導向程式設計，商品攝影及人像攝影。長期投入創客運動、資訊系統設計與開發、企業應用系統開發、軟體工程、新產品開發管理、商品及人像攝影等領域，並持續發表作品及相關專業著作。

Email:prgbruce@gmail.com

Line ID：dr.brucetsao

部落格：http://taiwanarduino.blogspot.tw/

書本範例網址：https://github.com/brucetsao/eDisplay

臉書社群(Arduino.Taiwan)：

https://www.facebook.com/groups/Arduino.Taiwan/

活動官網：http://www.accupass.com/org/detail/r/1604171414586144128670/1/0

Youtube：https://www.youtube.com/channel/UCcYG2yY_u0m1aotcA4hrRgQ

許智誠 (Chih-Cheng Hsu) ，美國加州大學洛杉磯分校(UCLA) 資訊工程系博士，曾任職於美國 IBM 等軟體公司多年，現任教於中央大學資訊管理學系專任副教授，主要研究為軟體工程、設計流程與自動化、數位教學、雲端裝置、多層式網頁系統、系統整合。

Email: khsu@mgt.ncu.edu.tw

蔡英德 (Yin-Te Tsai) ，國立清華大學資訊科學系博士，目前是靜宜大學資訊傳播工程學系教授、靜宜大學計算機及通訊中心主任，主要研究為演算法設計與分析、生物資訊、軟體開發、視障輔具設計與開發。

Email:yttsai@pu.edu.tw

附 錄

分壓線路

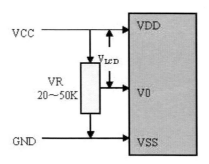

LCM 1602 原廠資料

 TC1602A-01T

FUNCTIONS & FEATURES

- Construction   : COB (Chip-on-Board)
- Display Format   : 16x2 Characters
- Display Type   : STN, Transflective, Positive, Y-G
- Controller   : SPLC780D1 or equivalent controller
- Interface   :   8-bit parallel interface
- Backlight   : yellow-green\bottom lights
- Viewing Direction   : 6 O'clock
- Driving Scheme   : 1/16 Duty Cycle, 1/5 Bias
- Power Supply Voltage   : 5.0 V
- V_{LCD} Adjustable For Best Contrast   : 5.0 V (V_{op}.)
- Operation temperature   : -10℃ to +60℃
- Storage temperature   : -20℃ to +70℃

BLOCK DIAGRAM

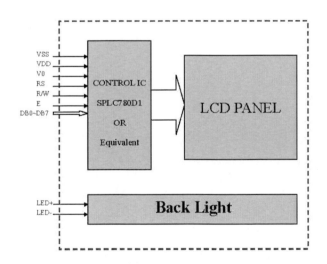

Ver.V00 2009-09-23   www.tinsharp.com

資料來源：(DFRobot, 2013; Guangzhou_Tinsharp_Industrial_Corp._Ltd., 2013)

MODULE OUTLINE DRAWING

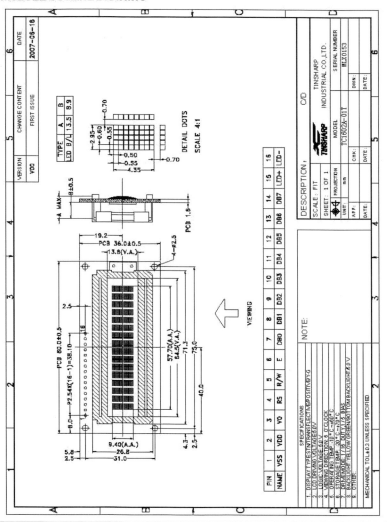

INTERFACE PIN FUNCTIONS

Pin No.	Symbol	Level	Description
1	VSS	0V	Ground.
2	VDD	+5.0V	Power supply for logic operating.
3	V0	--	Adjusting supply voltage for LCD driving.
4	RS	H/L	A signal for selecting registers: 1: Data Register (for read and write) 0: Instruction Register (for write), Busy flag-Address Counter (for read).
5	R/W	H/L	R/W = "H": Read mode. R/W = "L": Write mode.
6	E	H/L	An enable signal for writing or reading data.
7	DB0	H/L	
8	DB1	H/L	
9	DB2	H/L	
10	DB3	H/L	This is an 8-bit bi-directional data bus.
11	DB4	H/L	
12	DB5	H/L	
13	DB6	H/L	
14	DB7	H/L	
15	LED+	+5.0V	Power supply for backlight.
16	LED-	0V	The backlight ground.

ABSOLUTE MAXIMUM RATINGS (Ta = 25℃)

Parameter	Symbol	Min	Max	Unit
Supply voltage for logic	V_{DD}	-0.3	+7.0	V
Supply voltage for LCD	V_o	0	V_{DD}+0.3	V
Input voltage	V_I	-0.3	V_{DD}+0.3	V
Normal Operating temperature	T_{OP}	-20	+70	℃
Normal Storage temperature	T_{ST}	-30	+80	℃

Note: Stresses beyond those given in the Absolute Maximum Rating table may cause operational errors or damage to the device. For normal operational conditions see AC/DC Electrical Characteristics.

DC ELECTRICAL CHARACTERISTICS

Parameter	Symbol	Condition	Min	T_{YP}	Max	Unit
Supply voltage for logic	VDD	--	4.8	5.0	5.2	V
Supply current for logic	IDD	--	--	120	150	mA
Operating voltage for LCD	VLCD	-10℃				V
		25℃	4.8	5.0	5.2	
		+60℃				
Input voltage "H" level	VIH	--	0.7 VDD	--	VDD+0.3	V
Input voltage "L" level	VIL	--	0	--	0.2VDD	V

LED BACKLIGHT CHARACTERISTICS

COLOR	Wavelength λ p(nm)	Operating Voltage(±0.15V)	Spectral line half width Δλ (nm)	Forward Current (mA)
Yellow-green	--	4.1	--	100

NOTE: Do not connect +5V directly to the backlight terminals. This will ruin the backlight.

Ver.V00 2009-09-23

www.tinsharp.com

CONNECTION WITH MCU

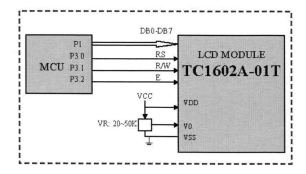

(1) Typical V0 connections for display contrast

Adjust V0 to +5.0V (VLCD=+5V) as an initial setting. When the module is operational, readjust V0 for optimal display appearance.

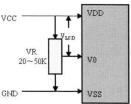

We recommend allowing field adjustment of V0 for all designs. The optimal value for V0 will change with temperature, variations in VDD, and viewing angle. V0 will also vary module-to-module and batch-to-batch due to normal manufacturing variations.

Ideally, adjustment to V0 should be available to the end user so each user can adjust the display to the optimal contrast for their required viewing conditions. As a minimum, your design should allow V0 to be adjusted as part of your product's final test.

Although a potentiometer is shown as a typical connection, V0 can be driven by your microcontroller, either by using a DAC or a filtered PWM. Displays that require V0 to be negative may need a level-shifting circuit. Please do not hesitate to contact Tinsharp application support for design assistance on your application.

(2) MPU Interface 4-bit/8-Bit

There are tow types of data operations: 4-bit and 8-bit operations. Using 4-bit MPU, the interfacing 4-bit data is transferred by 4-busline (DB4~DB7). Thus, DB0 to DB3 bus lines are not used. Using 4-bit MPU to interface 8-bit data requires tow times transferring. First, the higher 4-bit data is transferred by 4-busline (for 8-bit operation, DB7~DB4). Secondly, the lower 4-bit data is transferred by 4-busline (for 8-bit operation, DB3~DB0). For 8-bit MPU, the 8-bit data is transferred by 8-busline (DB0~DB7).

Ver.V00 2009-09-23

AC CHARACTERISTICS

(1) Write Mode (Writing data from MPU to SPLC780D1)

Characteristics	Symbol	Limit			Unit	Test Condition
		Min.	Typ.	Max.		
E Cycle Time	t_C	400	-	-	ns	Pin E
E Pulse Width	t_{PW}	150	-	-	ns	Pin E
E Rise/Fall Time	t_R, t_F	-	-	25	ns	Pin E
Address Setup Time	t_{SP1}	30	-	-	ns	Pins: RS, R/W, E
Address Hold Time	t_{HD1}	10	-	-	ns	Pins: RS, R/W, E
Data Setup Time	t_{SP2}	40	-	-	ns	Pins: DB0 - DB7
Data Hold Time	t_{HD2}	10	-	-	ns	Pins: DB0 - DB7

Write Mode Timing Diagram (Writing data from MPU to SPLC780D1)

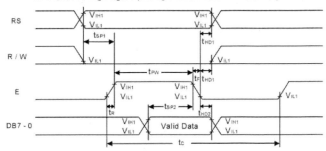

(2) Read Mode (Reading data from SPLC780D1 to MPU)

Characteristics	Symbol	Limit			Unit	Test Condition
		Min.	Typ.	Max.		
E Cycle Time	t_C	400	-	-	ns	Pin E
E Pulse Width	t_W	150	-	-	ns	Pin E
E Rise/Fall Time	t_R, t_F	-	-	25	ns	Pin E
Address Setup Time	t_{SP1}	30	-	-	ns	Pins: RS, R/W, E
Address Hold Time	t_{HD1}	10	-	-	ns	Pins: RS, R/W, E
Data Output Delay Time	t_D	-	-	100	ns	Pins: DB0 - DB7
Data hold time	t_{HD2}	5.0	-	-	ns	Pin DB0 - DB7

Read Mode Timing Diagram (Reading data from SPLC780D1 to MCU)

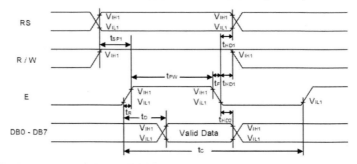

(3) Interface mode with LCD driver (SPLC100B1)

Characteristics	Symbol	Limit			Unit	Test Condition
		Min.	Typ.	Max.		
Clock pulse width high	t_{PWH}	800	-	-	ns	Pins. CL1, CL2
Clock pulse width low	t_{PWL}	800	-	-	ns	Pins. CL1, CL2
Clock setup time	t_{CSP}	500	-	-	ns	Pins. CL1, CL2
Data setup time	t_{DSP}	300	-	-	ns	Pins. D
Data hold time	t_{HD}	300	-	-	ns	Pins. D
M delay time	t_D	-1000	-	1000	ns	Pins. M

Interface mode with SPLC100B1 Timing Diagram

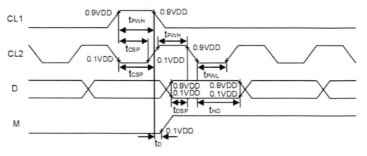

OPTICAL CHARACTERISTICS

ITEM	SYMBOL	CONDITION	MIN	TYP	MAX	UNIT	NOTE
Contrast ratio	CR	θ=0, Φ=0	-	3	-		
Response time(rise)	Tr	25℃		-	250	ms	
Response time(fall)	Td			-	350		
Viewing angle	θf	25℃				deg.	
	θb						
	θl		-				
	θr		-				

Note1: Definition Operation Voltage (V_OP)

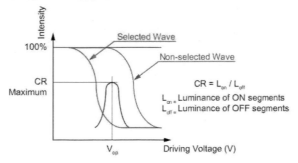

Note2: Response time

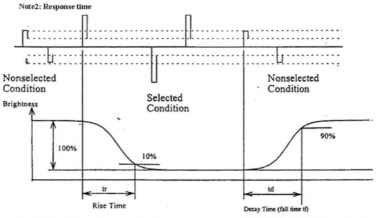

Note3: Viewing angle

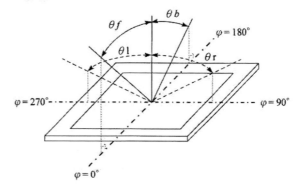

COMMAND TABLE

Instruction	Instruction Code										Description	Execution time (Temp = 25℃)		
	RS	RW	DB7	DB6	DB5	DB4	DB3	DB2	DB1	DB0		Fosc= 190KHz	Fosc= 270KHz	Fosc= 350KHz
Clear Display	0	0	0	0	0	0	0	0	0	1	Write "20H" to DDRAM and set DDRAM address to "00H" from AC	2.16ms	1.52ms	1.18ms
Return Home	0	0	0	0	0	0	0	0	1	-	Set DDRAM address to "00H" from AC and return cursor to its original position if shifted. The contents of DDRAM are not changed.	2.16ms	1.52ms	1.18ms
Entry Mode Set	0	0	0	0	0	0	0	1	I/D	S	Assign cursor moving direction and enable the shift of entire display	53μs	38μs	29μs
Display ON/ OFF Control	0	0	0	0	0	0	1	D	C	B	Set display (D), cursor(C), and blinking of cursor(B) on/off control bit.	53μs	38μs	29μs
Cursor or Display Shift	0	0	0	0	0	1	S/C	R/L	-	-	Set cursor moving and display shift control bit, and the direction, without changing of DDRAM data.	53μs	38μs	29μs
Function Set	0	0	0	0	1	DL	N	F	-	-	Set interface data length (DL: 8-bit/4-bit), numbers of display line (N: 2-line/1-line) and, display font type (F:5x10 dots/5x8 dots)	53μs	38μs	29μs
Set CGRAM Address	0	0	0	1	AC5	AC4	AC3	AC2	AC1	AC0	Set CGRAM address in address counter.	53μs	38μs	29μs
Set DDRAM Address	0	0	1	AC6	AC5	AC4	AC3	AC2	AC1	AC0	Set DDRAM address in address counter	53μs	38μs	29μs
Read Busy Flag and Address Counter	0	1	BF	AC6	AC5	AC4	AC3	AC2	AC1	AC0	Whether during internal operation or not can be known by reading BF. The contents of address counter can also be read.			
Write Data to RAM	1	0	D7	D6	D5	D4	D3	D2	D1	D0	Write data into internal RAM (DDRAM/CGRAM).	53μs	38μs	29μs
Read Data from RAM	1	1	D7	D6	D5	D4	D3	D2	D1	D0	Read data from internal RAM (DDRAM/CGRAM).	53μs	38μs	29μs

Note1: "-": don't care
Note2: In the operation condition under -20℃ ~ 75℃, the maximum execution time for majority of instruction sets is 100us, except two instructions, "Clear Display" and "Return Home", in which maximum execution time can take up to 4.1ms.

RESET FUNCTION

At power on, SPLC780D1 starts the internal auto-reset circuit and executes the initial instructions. The initial procedures are shown as follows.

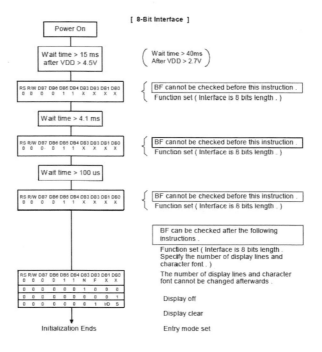

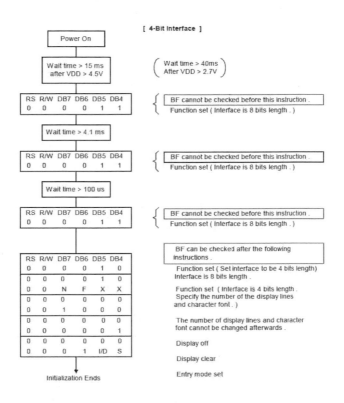

DISPLAY DATA RAM (DD RAM)

The 80-bit DD RAM is normally used for storing display data. Those DD RAM not used for display data can be used as general data RAM. Its address is configured in the Address Counter.

2 LINES X 16 CHARACTERS PER LINE																
Char.	1	2	3	4	5	6	7	8	9	10	11	12	13	14	15	16
Line 1	80	81	82	83	84	85	86	87	88	89	8A	8B	8C	8D	8E	8F
Line 2	C0	C1	C2	C3	C4	C5	C6	C7	C8	C9	CA	CB	CC	CD	CE	CF

Timing Generation Circuit

The timing generating circuit is able to generate timing signals to the internal circuits. In order to prevent the internal timing interface, the MPU access timing and the RAM access timing are generated independently.

LCD Driver Circuit

Total of 16 commons and 40 segments signal drivers are valid in the LCD driver circuit. When a program specifies the character fonts and line numbers, the corresponding common signals output drive-waveforms and the others still output unselected waveforms. The relationships between Display Data RAM Address and LCD's position are depicted as follows.

Character Generator ROM (CG ROM)

Using 8-bit character code, the character generator ROM generates 5 x 8 dots or 5 x 10 dots character patterns. It also can generate 192's 5 x 8 dots character patterns and 64's 5 x 10 dots character patterns.

Character Generator RAM (CG RAM)

Users can easily change the character patterns in the character generator RAM through program. It can be written to 5 x 8 dots, 8-character patterns or 5 x 10 dots for 4-character patterns.

CHARACTER GENERATOR ROM

SPLC780D1-001A:

Upper 4 bit / Lower 4 bit	LLLL	LLLH	LLHL	LLHH	LHLL	LHLH	LHHL	LHHH	HLLL	HLLH	HLHL	HLHH	HHLL	HHLH	HHHL	HHHH
LLLL			0	@	P	`	p					―	タ	ミ	α	p
LLLH			!	1	A	Q	a	q			。	ア	チ	ム	ä	q
LLHL			"	2	B	R	b	r			「	イ	ツ	メ	β	θ
LLHH			#	3	C	S	c	s			」	ウ	テ	モ	ε	∞
LHLL			$	4	D	T	d	t			、	エ	ト	ヤ	μ	Ω
LHLH			%	5	E	U	e	u			・	オ	ナ	ユ	σ	ü
LHHL			&	6	F	V	f	v			ヲ	カ	ニ	ヨ	ρ	Σ
LHHH			'	7	G	W	g	w			ア	キ	ヌ	ラ	g	π
HLLL			(	8	H	X	h	x			イ	ク	ネ	リ	√	x̄
HLLH			)	9	I	Y	i	y			ゥ	ケ	ノ	ル	‐	y
HLHL			*	:	J	Z	j	z			エ	コ	ハ	レ	j	千
HLHH			+	;	K	[	k	{			オ	サ	ヒ	ロ	x	万
HHLL			,	<	L	¥	l	|			ャ	シ	フ	ワ	¢	円
HHLH			-	=	M	]	m	}			ュ	ス	ヘ	ン	±	÷
HHHL			.	>	N	^	n	→			ョ	セ	ホ	゛	ñ	
HHHH			/	?	O	_	o	←			ッ	ソ	マ	゜	ö	■

RELIABILITY TEST CONDITION

No.	TEST Item	Content of Test	Test Condition	Applicable Standard
1	High temperature storage	Endurance test applying the high storage Temperature for a long time.	70° C 96hrs	------
2	Low temperature storage	Endurance test applying the low storage Temperature for a long time	-20° C 96hrs	-----
3	High temperature operation	Endurance test applying the electric stress (Voltage & current)and the thermal stress to the element for a long time	60° C 96hrs	------
4	Low temperature operation	Endurance test applying the electric stress Under low temperature for a long time	-10° C 96hrs	------
5	High temperature/ Humidity storage	Endurance test applying the electric stress(Voltage & current) and Temperature/ Humidity stress to the element for a long time	40° C 90%RH 96hrs	
6	High temperature/ Humidity operation	Endurance test applying the electric stress (voltage & current)and temperature/ humidity stress to the element for a long time	40° C 90%RH 96hrs	
7	Temperature cycle	Endurance test applying the low and high temperature cycle. -10° C →25° C→60° C 30min←5min←30min.(1 cycle)	-10° C/60° C 10 cycle	------

Supply voltage for logic system = 5V. Supply voltage for LCD system = Operating voltage at 25° C.

Mechanical Test

Vibration test	Endurance test applying the vibration during transportation and using	10~22Hz→1.5mmp-p 22~500Hz→1.5G Total 0.5hour	
Shock test	Constructional and mechanical endurance test applying the shock during transportation.	50G half sign wave 11 msede 3 times of each direction	
Atmospheric pressure test	Endurance test applying the atmospheric pressure during transportation by air	115mbar 40hrs	
Static electricity test	Endurance test applying the electric stress to the terminal	VS=800V,RS-1.5K Ω CS=100pF, 1 time	

Environmental condition

The inspection should be performed at the 1metre height from the LCD module under 2 pieces of 40W white fluorescent lamps (Normal temperature 20~25℃ and normal humidity 60±15%RH).

Ver.V00 2009-09-23   *www.tinsharp.com*

PRECAUTION FOR USING LCM MODULE

- Please remove the protection foil of polarizer before using.

- The display panel is made of glass. Do not subject it to a mechanical shock by dropping it from a high place, etc.

- If the display panel is damaged and the liquid crystal substance inside it leaks out, do not get any in your mouth. If the substance come into contact with your skin or clothes promptly wash it off using soap and water.

- Do not apply excessive force to the display surface or the adjoining areas since this may cause the color tone to vary.

- The polarizer covering the display surface of the LCD module is soft and easily scratched. Handle this polarize carefully.

- To prevent destruction of the elements by static electricity, be careful to maintain an optimum work environment.
  - Be sure to ground the body when handling the LCD module.
  - Tools required for assembly, such as soldering irons, must be properly grounded.
  - To reduce the amount of static electricity generated, do not conduct assembly and other work under dry conditions.
  - The LCD module is coated with a film to protect the display surface. Exercise care when peeling off this protective film since static electricity may be generated.

- Storage precautions
  When storing the LCD modules, avoid exposure to direct sunlight or to the light of fluorescent lamps.
  Keep the modules in bags designed to prevent static electricity charging under low temperature / normal humidity conditions (avoid high temperature / high humidity and low temperatures below 0℃).Whenever possible, the LCD modules should be stored in the same conditions in which they were shipped from our company.

OTHERS

- Liquid crystals solidify at low temperature (below the storage temperature range) leading to defective orientation of liquid crystal or the generation of air bubbles (black or white). Air bubbles may also be generated if the module is subjected to a strong shock at a low temperature.

- If the LCD modules have been operating for a long time showing the same display patterns may remain on the screen as ghost images and a slight contrast irregularity may also appear. Abnormal operating status can be resumed to be normal condition by suspending use for some time. It should be noted that this phenomena does not adversely affect performance reliability.

- To minimize the performance degradation of the LCD modules resulting from caused by static electricity, etc. exercise care to avoid holding the following   sections when handling the modules :
  - Exposed area of the printed circuit board
  - Terminal electrode sections

A. DATE CODE RULES

A.1. DATE CODE FOR SAMPLE

YP: meaning sample

TC1602A-01T ——→ LCM part number
YP XXXXX ——→ Sample array No.

A.2. DATE CODE FOR PRODUCTION

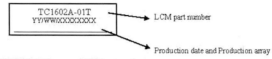

TC1602A-01T ——→ LCM part number
YY/WW/XXXXXXXX

——→ Production date and Production array

A. TC1602A-01T represents LCM part number

C. YY/WW represents Year, Week

    YY—Year   WW—Week

    XXXXXXXX—Production array No.

B. CHANGE NOTES:

Ver.	Descriptions	Editor	Date
V00	First Issue	HXY	2009-09-23

LCM 2004 原廠資料

SHENZHEN EONE ELECTRONICS CO.,LTD

Specification

For

LCD Module

2004A

2004A LCD Module Specification Ver1.0 Total 14 pages1

TABLE OF CONTENTS

1.0 INTRODUCTION

This USER'S MANUAL is introduced the outside dimensions, optical characteristics, electrical characteristics, interface, controller commands, etc. of the custom design LCD module.

1.1 FEATURE

(1) Display mode: STN POSITIVE, TRANSFLECTIVE, YELLOW-GREEN COLOR

(2) Display format: 20 characters X 4 line

(3) Driving method: 1/16 Duty, 1/5 Bias

(4) Viewing direction: 6 o'clock

(5) Control IC: SPLC780D

(6) Interface Input Data : 4-Bits or 8-Bits interface available

(7) Back light: LED (Yellow-Green)

2.0 DIMENSION DIAGRAM

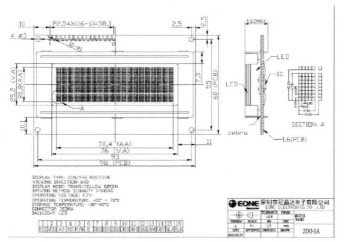

2004A   LCD Module Specification   Ver1.1   Total 14   pages3

3.0 MECHANICAL SPECIFICATIONS

ITEM	STANDARD VALUE	UNIT
DOTS	5X8	characters -
DOT SIZE	0.55X0.55	mm
DOT PITCH	0.60X0.60	mm
MODULE DIMENSION	98.0(W) × 60.0(H) × 1.6(T)	mm
EFFECTIVE DISPLAY AREA	84.0(W) × 31.0(H)MIN	mm

4.0 MAX STANDARD VALUE

ITEM	SYMBOL	MIN.	TYPE	MAX	UNIT
OPERATING TEMPERATURE	Top	-10	25	60	℃
STORAGE TEMPERATURE	Tst	-20	/	70	℃
INPUT VOLTAGE	VI	VSS	/	VDD	V
SUPPLY VOLTAGE FOR LOGIC	VDD-VSS	-0.3	/	7.0	V
SUPPLY VOLTAGE FOR LCD	VDD-V0	VDD-10.0	/	VDD+0.3	V

5.0 ELECTRICAL CHARACTERISTICS

ITEM	SYMBOL	CONDITION	MIN.	TYP.	MAX.	UNIT

SUPPLY VOLTAGE FOR LOGIC	V_{DD}-V_{SS}	Ta = 25 °C	2.7	5.0	5.5	V
SUPPLY VOLTAGE FOR LCD	V_{DD}-V_O (VOP)	Ta = 25 °C	3.0	5.0	10.0	V
INPUT HIGH VOL.	V_{IH}	Ta = 25 °C	0.7VDD	-	VDD	V
INPUT LOW VOL.	V_{IL}	Ta = 25 °C	-0.3	-	0.6	V
OUTPUT HIGH VOL.	V_{OH}	Ta = 25 °C	0.75VDD	-	-	V
OUTPUT LOW VOL.	V_{OL}	Ta = 25 °C	-	-	0.2VDD	V
SUPPLY CURRENT	IDD	V_{DD} = 3.0V	-	0.1	0.25	mA

6.0 OPTICAL CHARACTERISTICS

No	Item		Symbol	Measurement temperature	MIN.	TYP.	MAX.	Unit
1	Contrast Ratio		Cr	25℃	2.60	3.17		
2	Response Time	Rise time	Tr	25℃	-	-	0.2	us
		Fall time	Tf	25℃	-	-	0.2	us
3	Viewing Angle	6H,Φ=0°	θ 1	25℃	55			Deg.
		12H,Φ=180	θ 2	25℃	0			Deg.
		Φ=90°	θ 3	25℃	45			Deg.
		Φ=270	θ 4	25℃	45			Deg.
4	Frame Frequency			25℃	190	270	350	Hz

6.1 OPTICAL MEASUREMENT SYSTEM

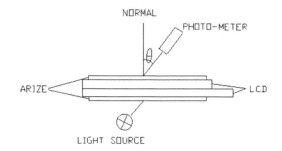

6.2 DEFINITION OF θ AND Φ

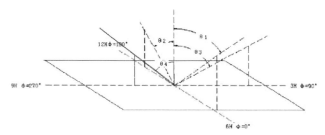

6.3 DEFINITION OF CONTRAST RATIO Cr

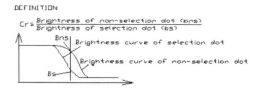

6.4 DEFINITION OF OPTICAL RESPONSE TIME

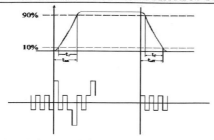

7.0 INTERFACE PIN FUNCTION DESCRIPTION

PIN NO	SYMBOL	FUNCTION
1	VSS	Power Ground
2	VDD	Power supply for logic circuit (+5V)
3	V0	For LCD drive voltage (variable)
4	RS (C/D)	H: Display Data, L:Display Instruction
5	R/W	H: Data Read (LCM to MPU) ; L: Data Write (MPU to LCM)
6	E	Enable signal. Write mode (R/W = L) data of DB<0:7> is latched at the falling edge of E. Read mode (R/W = H) DB<0:7> appears the reading data while E is at high level
7-14	DB0-DB7	Data bus. There state I/O common terminal.
15	A	Power for LED Backlight(+5V)
16	K	Power for LED Backlight(Ground)

8.0 BLOCK DIAGRAM

2004A LCD Module Specification   Ver1.1   Total 14   pages7

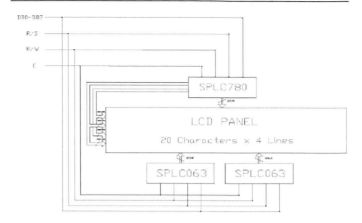

8.1 POWER SUPPLY BLOCK DIAGRAM

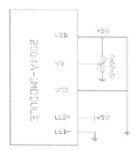

9.0 TIMING CHARACTERISTICS

7.3.3. Write mode (Writing data from MPU to SPLC780D)

Characteristics	Symbol	Limit			Unit	Test Condition
		Min.	Typ.	Max.		
E Cycle Time	t_C	1000	-	-	ns	Pin E
E Pulse Width	t_{PW}	450	-	-	ns	Pin E
E Rise/Fall Time	t_r, t_f	-	-	25	ns	Pin E
Address Setup Time	t_{SP1}	60	-	-	ns	Pins: RS, R/W, E
Address Hold Time	t_{HD1}	20	-	-	ns	Pins: RS, R/W, E
Data Setup Time	t_{SP2}	195	-	-	ns	Pins: DB0 - DB7
Data Hold Time	t_{HD2}	10	-	-	ns	Pins: DB0 - DB7

7.3.4. Read mode (Reading data from SPLC780D to MPU)

Characteristics	Symbol	Limit			Unit	Test Condition
		Min.	Typ.	Max.		
E Cycle Time	t_C	1000	-	-	ns	Pin E
E Pulse Width	t_{PW}	450	-	-	ns	Pin E
E Rise/Fall Time	t_r, t_f	-	-	25	ns	Pin E
Address Setup Time	t_{SP1}	60	-	-	ns	Pins: RS, R/W, E
Address Hold Time	t_{HD1}	20	-	-	ns	Pins: RS, R/W, E
Data Output Delay Time	t_D	-	-	360	ns	Pins: DB0 - DB7
Data hold time	t_{HD2}	5.0	-	-	ns	Pin DB0 - DB7

7.5.6. Write mode timing diagram (Writing Data from MPU to SPLC780D)

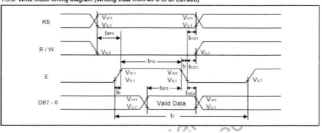

7.5.7. Read mode timing diagram (Reading Data from SPLC780D to MPU)

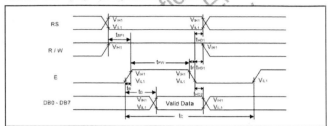

7.5.8. Interface mode with SPLC100A1 timing diagram

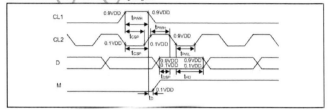

10.0 Display control instruction

The display control instructions control the internal state of the SPLC780D-01. Instruction is received from

MPU to SPLC780D-01for the display control. The following table shows various instructions.

6.3. Instruction Table

Instruction	Instruction Code										Description	Execution time (Temp = 25℃)		
	RS	RW	DB7	DB6	DB5	DB4	DB3	DB2	DB1	DB0		Fosc= 190KHz	Fosc= 270KHz	Fosc= 350KHz
Clear Display	0	0	0	0	0	0	0	0	0	1	Write "20H" to DDRAM and set DDRAM address to "00H" from AC	2.16ms	1.52ms	1.18ms
Return Home	0	0	0	0	0	0	0	0	1	-	Set DDRAM address to "00H" from AC and return cursor to its original position if shifted. The contents of DDRAM are not changed.	2.16ms	1.52ms	1.18ms
Entry Mode Set	0	0	0	0	0	0	0	1	I/D	S	Assign cursor moving direction and enable the shift of entire display	53μs	38μs	29μs
Display ON/ OFF Control	0	0	0	0	0	0	1	D	C	B	Set display (D), cursor(C), and blinking of cursor(B) on/off control bit.	53μs	38μs	29μs
Cursor or Display Shift	0	0	0	0	0	1	S/C	R/L	-	-	Set cursor moving and display shift control bit, and the direction, without changing of DDRAM data.	53μs	38μs	29μs
Function Set	0	0	0	0	1	DL	N	F	-	-	Set interface data length (DL: 8-bit/4-bit), numbers of display line (N: 2-line/1-line) and, display font type (F:5x10 dots/5x8 dots)	53μs	38μs	29μs
Set CGRAM Address	0	0	0	1	AC5	AC4	AC3	AC2	AC1	AC0	Set CGRAM address in address counter.	53μs	38μs	29μs
Set DDRAM Address	0	0	1	AC6	AC5	AC4	AC3	AC2	AC1	AC0	Set DDRAM address in address counter	53μs	38μs	29μs
Read Busy Flag and Address Counter	0	1	BF	AC6	AC5	AC4	AC3	AC2	AC1	AC0	Whether during internal operation or not can be known by reading BF. The contents of address counter can also be read.			
Write Data to RAM	1	0	D7	D6	D5	D4	D3	D2	D1	D0	Write data into internal RAM (DDRAM/CGRAM).	53μs	38μs	29μs
Read Data from RAM	1	1	D7	D6	D5	D4	D3	D2	D1	D0	Read data from internal RAM (DDRAM/CGRAM).	53μs	38μs	29μs

Note1: "-": don't care

Note2: In the operation condition under -20℃ ~ 75℃, the maximum execution time for majority of instruction sets is 100us, except two instructions, "Clear Display" and "Return Home", in which maximum execution time can take up to 4.1ms.

■ Instruction Description

● **Clear Display**

RS RW DB7 DB6 DB5 DB4 DB3 DB2 DB1 DB0

Code	0	0	0	0	0	0	0	0	0	1

Clear all the display data by writing "20H" (space code) to all DDRAM address, and set DDRAM address to "00H" into AC (address counter). Return cursor to the original status, namely, bring the cursor to the left edge on first line of the display. Make entry mode increment (I/D = "1").

● **Return Home**

RS RW DB7 DB6 DB5 DB4 DB3 DB2 DB1 DB0

Code	0	0	0	0	0	0	0	0	1	x

Return Home is cursor return home instruction. Set DDRAM address to "00H" into the address counter. Return cursor to its original site and return display to its original status, if shifted. Contents of DDRAM does not change.

● **Entry Mode Set**

RS RW DB7 DB6 DB5 DB4 DB3 DB2 DB1 DB0

Code	0	0	0	0	0	0	0	1	I/D	S

Set the moving direction of cursor and display.

➢ **I/D : Increment / decrement of DDRAM address (cursor or blink)**
When I/D = "High", cursor/blink moves to right and DDRAM address is increased by 1.
When I/D = "Low", cursor/blink moves to left and DDRAM address is decreased by 1.
* CGRAM operates the same as DDRAM, when read from or write to CGRAM.

➢ **S: Shift of entire display**
When DDRAM read (CGRAM read/write) operation or S = "Low", shift of entire display is not performed. If S = "High" and DDRAM write operation, shift of entire display is performed according to I/D value (I/D = "1" : shift left, I/D = "0" : shift right).

S	I/D	Description
H	H	Shift the display to the left
H	L	Shift the display to the right

- **Display ON/OFF**

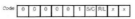

RS RW DB7 DB6 DB5 DB4 DB3 DB2 DB1 DB0

Code	0	0	0	0	0	0	1	D	C	B

Control display/cursor/blink ON/OFF 1 bit register.

- **D : Display ON/OFF control bit**
  When D = "High", entire display is turned on.

  When D = "Low", display is turned off, but display data is remained in DDRAM.
- **C : Cursor ON/OFF control bit**
  When C = "High", cursor is turned on.

  When C = "Low", cursor is disappeared in current display, but I/D register remains its data.
- **B : Cursor Blink ON/OFF control bit**
  When B = "High", cursor blink is on, that performs alternate between all the high data and display

  character at the cursor position.

  When B = "Low", blink is off.

- **Cursor or Display Shift**

RS RW DB7 DB6 DB5 DB4 DB3 DB2 DB1 DB0

Code	0	0	0	0	0	1	S/C	R/L	x	x

Without writing or reading of display data, shift right/left cursor position or display. This instruction is used to correct or search display data. During 2-line mode display, cursor moves to the 2nd line after 40th digit of 1st line. Note that display shift is performed simultaneously in all the line. When displayed data is shifted repeatedly, each line shifted individually. When display shift is performed, the contents of address counter are not changed.

S/C	R/L	Description	AC Value
L	L	Shift cursor to the left	AC=AC-1
L	H	Shift cursor to the right	AC=AC+1
H	L	Shift display to the left. Cursor follows the display shift	AC=AC
H	H	Shift display to the right. Cursor follows the display shift	AC=AC

- **Function Set**

RS RW DB7 DB6 DB5 DB4 DB3 DB2 DB1 DB0

Code	0	0	0	0	1	DL	N	F	x	x

➤ **DL : Interface data length control bit**
When DL = "High", it means 8-bit bus mode with MPU.

When DL = "Low", it means 4-bit bus mode with MPU. So to speak, DL is a signal to select 8-bit or 4-bit bus mode.

When 4-bit bus mode, it needs to transfer 4-bit data by two times.

➤ **N : Display line number control bit**
When N = "Low", it means 1-line display mode.

When N = "High", 2-line display mode is set.

➤ **F : Display font type control bit**
When F = "Low", it means 5 x 8 dots format display mode

When F = "High", 5 x11 dots format display mode.

N	F	No. of Display Lines	Character Font	Duty Factor
L	L	1	5x8	1/8
L	H	1	5x11	1/11
H	x	2	5x8	1/16

● **Set CGRAM Address**

Set CGRAM address to AC.

This instruction makes CGRAM data available from MPU.

● **Set DDRAM Address**

Code	RS	RW	DB7	DB6	DB5	DB4	DB3	DB2	DB1	DB0
	0	0	1	AC6	AC5	AC4	AC3	AC2	AC1	AC0

Set DDRAM address to AC.

This instruction makes DDRAM data available from MPU.

When 1-line display mode (N = 0), DDRAM address is from "00H" to "4FH".

In 2-line display mode (N = 1), DDRAM address in the 1st line is from "00H" to "27H", and DDRAM address in the 2nd line is from "40H" to "67H".

● **Read Busy Flag and Address**

Code	RS	RW	DB7	DB6	DB5	DB4	DB3	DB2	DB1	DB0
	0	1	BF	AC6	AC5	AC4	AC3	AC2	AC1	AC0

When BF = "High", indicates that the internal operation is being processed.So during this time the next instruction cannot be accepted.

The address Counter (AC) stores DDRAM/CGRAM addresses, transferred from IR.

After writing into (reading from) DDRAM/CGRAM, AC is automatically increased (decreased) by 1.

● **Write Data to CGRAM or DDRAM**

	RS	RW	DB7	DB6	DB5	DB4	DB3	DB2	DB1	DB0
Code	1	0	D7	D6	D5	D4	D3	D2	D1	D0

Write binary 8-bit data to DDRAM/CGRAM.

The selection of RAM from DDRAM, CGRAM, is set by the previous address set instruction : DDRAM address set, CGRAM address set. RAM set instruction can also determine the AC direction to RAM.

After write operation, the address is automatically increased/decreased by 1, according to the entry mode.

● **Read Data from CGRAM or DDRAM**

	RS	RW	DB7	DB6	DB5	DB4	DB3	DB2	DB1	DB0
Code	1	1	D7	D6	D5	D4	D3	D2	D1	D0

Read binary 8-bit data from DDRAM/CGRAM.

The selection of RAM is set by the previous address set instruction. If address set instruction of RAM is not performed before this instruction, the data that read first is invalid, because the direction of AC is not determined. If you read RAM data several times without RAM address set instruction before read operation, you can get correct RAM data from the second, but the first data would be incorrect, because there is no time margin to transfer RAM data.

In case of DDRAM read operation, cursor shift instruction plays the same role as DDRAM address set instruction : it also transfer RAM data to output data register. After read operation address counter is automatically increased/decreased by 1 according to the entry mode. After CGRAM read operation, display shift may not be executed correctly.

* In case of RAM write operation, after this AC is increased/decreased by 1 like read operation. In this time, AC indicates the next address position, but you can read only the previous data by read instruction.

Max7219 Led 模組線路圖

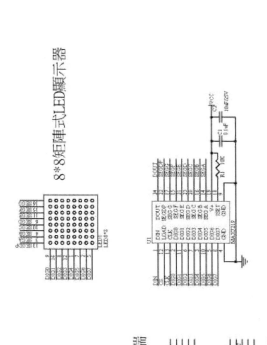

MAX7219 資料

MAX7219/MAX7221

Serially Interfaced, 8-Digit LED Display Drivers

General Description

The MAX7219/MAX7221 are compact, serial input/output common-cathode display drivers that interface microprocessors (µPs) to 7-segment numeric LED displays of up to 8 digits, bar-graph displays, or 64 individual LEDs. Included on-chip are a BCD code-B decoder, multiplex scan circuitry, segment and digit drivers, and an 8x8 static RAM that stores each digit. Only one external resistor is required to set the segment current for all LEDs. The MAX7221 is compatible with SPI™, QSPI™, and MICROWIRE™, and has slew-rate-limited segment drivers to reduce EMI.

A convenient 4-wire serial interface connects to all common µPs. Individual digits may be addressed and updated without rewriting the entire display. The MAX7219/MAX7221 also allow the user to select code-B decode or no-decode for each digit.

The devices include a 150µA low-power shutdown mode, analog and digital brightness control, a scan-limit register that allows the user to display from 1 to 8 digits, and a test mode that forces all LEDs on.

For applications requiring 3V operation or segment blinking, refer to the MAX6951 data sheet.

Applications

Bar-Graph Displays

Industrial Controllers

Panel Meters

LED Matrix Displays

Features

♦ 10MHz Serial Interface

♦ Individual LED Segment Control

♦ Decode/No-Decode Digit Selection

♦ 150µA Low-Power Shutdown (Data Retained)

♦ Digital and Analog Brightness Control

♦ Display Blanked on Power-Up

♦ Drive Common-Cathode LED Display

♦ Slew-Rate Limited Segment Drivers
  for Lower EMI (MAX7221)

♦ SPI, QSPI, MICROWIRE Serial Interface (MAX7221)

♦ 24-Pin DIP and SO Packages

Ordering Information

PART	TEMP RANGE	PIN-PACKAGE
MAX7219CNG	0°C to +70°C	24 Narrow Plastic DIP
MAX7219CWG	0°C to +70°C	24 Wide SO
MAX7219C/D	0°C to +70°C	Dice*
MAX7219ENG	-40°C to +85°C	24 Narrow Plastic DIP
MAX7219EWG	-40°C to +85°C	24 Wide SO
MAX7219ERG	-40°C to +85°C	24 Narrow CERDIP

Ordering Information continued at end of data sheet.
Dice are specified at T_A = +25°C.

Pin Configuration

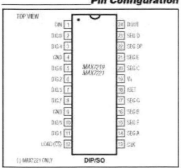

(*) MAX7221 ONLY   DIP/SO

Typical Application Circuit

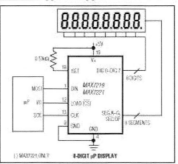

(*) MAX7221 ONLY   8-DIGIT µP DISPLAY

SPI and QSPI are trademarks of Motorola Inc. MICROWIRE is a trademark of National Semiconductor Corp.

For pricing, delivery, and ordering information, please contact Maxim Direct
at 1-888-629-4642, or visit Maxim's website at www.maximintegrated.com.   19-4452; Rev 4; 7/03

MAX7219/MAX7221
Serially Interfaced, 8-Digit LED Display Drivers

ABSOLUTE MAXIMUM RATINGS

Voltage (with respect to GND)
V+ .... -0.3V to 6V
DIN, CLK, LOAD, CS .... -0.3V to 6V
All Other Pins .... -0.3V to (V+ + 0.3V)
Current
DIG 0–DIG 7 Sink Current 500mA
SEG A–G, DP Source Current 100mA
Continuous Power Dissipation (T_A = +55°C)
Narrow Plastic DIP (derate 13.3mW/°C
above +70°C) 1066mW
Wide SO (derate 11.8mW/°C above +70°C) 941mW
Narrow CERDIP (derate 12.5mW/°C above +70°C) ...1000mW

Operating Temperature Ranges (T_{MIN} to T_{MAX})
MAX7219C_G/MAX7221C_G .... 0°C to +70°C
MAX7219E_G/MAX7221E_G .... -40°C to +85°C
Storage Temperature Range .... -65°C to +160°C
Lead Temperature (soldering, 10s) .... +300°C

Stresses beyond those listed under "Absolute Maximum Ratings" may cause permanent damage to the device. These are stress ratings only, and functional operation of the device at these or any other conditions beyond those indicated in the operational sections of the specifications is not implied. Exposure to absolute maximum rating conditions for extended periods may affect device reliability.

ELECTRICAL CHARACTERISTICS

(V+ = 5V ±10%, R_{SET} = 9.53kΩ ±1%, T_A = T_{MIN} to T_{MAX}, unless otherwise noted.)

PARAMETER	SYMBOL	CONDITIONS	MIN	TYP	MAX	UNITS
Operating Supply Voltage	V+		4.0		5.5	V
Shutdown Supply Current	I+	All digital inputs at V+ or GND, T_A = +25°C			150	μA
Operating Supply Current	I+	R_{SET} = open circuit			8	mA
		All segments and decimal point on, I_{SEG} = -40mA		330		
Display Scan Rate	f_{OSC}	8 digits scanned	500	800	1300	Hz
Digit Drive Sink Current	I_{DIGIT}	V+ = 5V, V_{OUT} = 0.65V	320			mA
Segment Drive Source Current	I_{SEG}	T_A = +25°C, V+ = 5V, V_{OUT} = (V+ - 1V)	-30	-40	-45	mA
Segment Current Slew Rate (MAX7221 only)	$\Delta I_{SEG}/\Delta t$	T_A = +25°C, V+ = 5V, V_{OUT} = (V+ - 1V)	10	20	50	mA/μs
Segment Drive Current Matching	ΔI_{SEG}			3.0		%
Digit Drive Leakage (MAX7221 only)	I_{DIGIT}	Digit off, V_{DIGIT} = V+			-10	μA
Segment Drive Leakage (MAX7221 only)	I_{SEG}	Segment off, V_{SEG} = 0V			1	μA
Digit Drive Source Current (MAX7219 only)	I_{DIGIT}	Digit off, V_{DIGIT} = (V+ - 0.3V)	-2			mA
Segment Drive Sink Current (MAX7219 only)	I_{SEG}	Segment off, V_{SEG} = 0.3V	5			mA

ELECTRICAL CHARACTERISTICS (continued)

(V+ = 5V ±10%, R$_{SET}$ = 9.53kΩ ±1%, T$_A$ = T$_{MIN}$ to T$_{MAX}$, unless otherwise noted.)

PARAMETER	SYMBOL	CONDITIONS	MIN	TYP	MAX	UNITS
LOGIC INPUTS						
Input Current DIN, CLK, LOAD, CS	I$_{IH}$, I$_{IL}$	V$_{IN}$ = 0V or V+	-1		1	μA
Logic High Input Voltage	V$_{IH}$		3.5			V
Logic Low Input Voltage	V$_{IL}$				0.8	V
Output High Voltage	V$_{OH}$	DOUT, I$_{SOURCE}$ = -1mA	V+ - 1			V
Output Low Voltage	V$_{OL}$	DOUT, I$_{SINK}$ = 1.6mA			0.4	V
Hysteresis Voltage	ΔV$_I$	DIN, CLK, LOAD, CS		1		V
TIMING CHARACTERISTICS						
CLK Clock Period	t$_{CP}$		100			ns
CLK Pulse Width High	t$_{CH}$		50			ns
CLK Pulse Width Low	t$_{CL}$		50			ns
CS Fall to SCLK Rise Setup Time (MAX7221 only)	t$_{CSS}$		25			ns
CLK Rise to CS or LOAD Rise Hold Time	t$_{CSH}$		0			ns
DIN Setup Time	t$_{DS}$		25			ns
DIN Hold Time	t$_{DH}$		0			ns
Output Data Propagation Delay	t$_{DO}$	C$_{LOAD}$ = 50pF			25	ns
Load-Rising Edge to Next Clock Rising Edge (MAX7219 only)	t$_{LDCK}$		50			ns
Minimum CS or LOAD Pulse High	t$_{CSW}$		50			ns
Data-to-Segment Delay	t$_{DSPD}$				2.25	ms

MAX7219/MAX7221
Serially Interfaced, 8-Digit LED Display Drivers

_____**Typical Operating Characteristics**

(V+ = +5V, TA = +25°C, unless otherwise noted.)

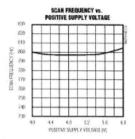

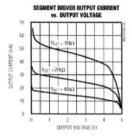

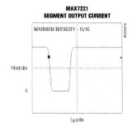

x

4

_____Pin Description

PIN	NAME	FUNCTION
1	DIN	Serial-Data Input. Data is loaded into the internal 16-bit shift register on CLK's rising edge.
2, 3, 5-8, 10, 11	DIG 0-DIG 7	Eight-Digit Drive Lines that sink current from the display common cathode. The MAX7219 pulls the digit outputs to V+ when turned off. The MAX7221's digit drivers are high-impedance when turned off.
4, 9	GND	Ground (both GND pins must be connected)
12	LOAD (MAX7219)	Load-Data Input. The last 16 bits of serial data are latched on LOAD's rising edge.
	CS (MAX7221)	Chip-Select Input. Serial data is loaded into the shift register while CS is low. The last 16 bits of serial data are latched on CS's rising edge.
13	CLK	Serial-Clock Input. 10MHz maximum rate. On CLK's rising edge, data is shifted into the internal shift register. On CLK's falling edge, data is clocked out of DOUT. On the MAX7221, the CLK input is active only while CS is low.
14-17, 20-23	SEG A-SEG G, DP	Seven Segment Drives and Decimal Point Drive that source current to the display. On the MAX7219, when a segment driver is turned off it is pulled to GND. The MAX7221 segment drivers are high-impedance when turned off.
18	ISET	Connect to VDD through a resistor (RSET) to set the peak segment current (Refer to Selecting RSET Resistor and Using External Drivers section).
19	V+	Positive Supply Voltage. Connect to +5V.
24	DOUT	Serial-Data Output. The data into DIN is valid at DOUT 16.5 clock cycles later. This pin is used to daisy-chain several MAX7219/MAX7221's and is never high-impedance.

_____Functional Diagram

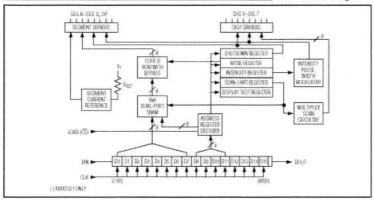

MAX7219/MAX7221
Serially Interfaced, 8-Digit LED Display Drivers

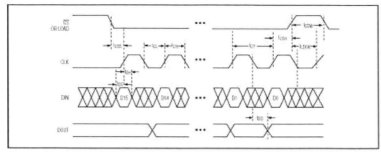

Figure 1. Timing Diagram

Table 1. Serial-Data Format (16 Bits)

D15	D14	D13	D12	D11	D10	D9	D8	D7	D6	D5	D4	D3	D2	D1	D0
X	X	X	X		ADDRESS			MSB			DATA				LSB

Detailed Description

MAX7219/MAX7221 Differences
The MAX7219 and MAX7221 are identical except for two parameters: the MAX7221 segment drivers are slew-rate limited to reduce electromagnetic interference (EMI), and its serial interface is fully SPI compatible.

Serial-Addressing Modes
For the MAX7219, serial data at DIN, sent in 16-bit packets, is shifted into the internal 16-bit shift register with each rising edge of CLK regardless of the state of LOAD. For the MAX7221, CS must be low to clock data in or out. The data is then latched into either the digit or control registers on the rising edge of LOAD/CS. LOAD/CS must go high concurrently with or after the 16th rising clock edge, but before the next rising clock edge or data will be lost. Data at DIN is propagated through the shift register and appears at DOUT 16.5 clock cycles later. Data is clocked out on the falling edge of CLK. Data bits are labeled D0-D15 (Table 1). D8-D11 contain the register address. D0-D7 contain the data, and D12-D15 are "don't care" bits. The first received is D15, the most significant bit (MSB).

Digit and Control Registers
Table 2 lists the 14 addressable digit and control registers. The digit registers are realized with an on-chip, 8x8 dual-port SRAM. They are addressed directly so that individual digits can be updated and retain data as long as V+ typically exceeds 2V. The control registers consist of decode mode, display intensity, scan limit (number of scanned digits), shutdown, and display test (all LEDs on).

Shutdown Mode
When the MAX7219 is in shutdown mode, the scan oscillator is halted, all segment current sources are pulled to ground, and all digit drivers are pulled to V+, thereby blanking the display. The MAX7221 is identical, except the drivers are high-impedance. Data in the digit and control registers remains unaltered. Shutdown can be used to save power or as an alarm to flash the display by successively entering and leaving shutdown mode. For minimum supply current in shutdown mode, logic inputs should be at ground or V+ (CMOS-logic levels).

Typically, it takes less than 250µs for the MAX7219/MAX7221 to leave shutdown mode. The display driver can be programmed while in shutdown mode, and shutdown mode can be overridden by the display-test function.

Table 2. Register Address Map

REGISTER	ADDRESS					HEX CODE
	D15–D12	D11	D10	D9	D8	
No-Op	X	0	0	0	0	0xX0
Digit 0	X	0	0	0	1	0xX1
Digit 1	X	0	0	1	0	0xX2
Digit 2	X	0	0	1	1	0xX3
Digit 3	X	0	1	0	0	0xX4
Digit 4	X	0	1	0	1	0xX5
Digit 5	X	0	1	1	0	0xX6
Digit 6	X	0	1	1	1	0xX7
Digit 7	X	1	0	0	0	0xX8
Decode Mode	X	1	0	0	1	0xX9
Intensity	X	1	0	1	0	0xXA
Scan Limit	X	1	0	1	1	0xXB
Shutdown	X	1	1	0	0	0xXC
Display Test	X	1	1	1	1	0xXF

Initial Power-Up

On initial power-up, all control registers are reset, the display is blanked, and the MAX7219/MAX7221 enter shutdown mode. Program the display driver prior to display use. Otherwise, it will initially be set to scan one digit, it will not decode data in the data registers, and the intensity register will be set to its minimum value.

Decode-Mode Register

The decode-mode register sets BCD code B (0-9, E, H, L, P, and -) or no-decode operation for each digit. Each bit in the register corresponds to one digit. A logic high selects code B decoding while logic low bypasses the decoder. Examples of the decode mode control-register format are shown in Table 4.

When the code B decode mode is used, the decoder looks only at the lower nibble of the data in the digit registers (D3–D0), disregarding bits D4–D6. D7, which sets the decimal point (SEG DP), is independent of the decoder and is positive logic (D7 = 1 turns the decimal point on). Table 5 lists the code B font.

When no-decode is selected, data bits D7-D0 correspond to the segment lines of the MAX7219/MAX7221. Table 6 shows the one-to-one pairing of each data bit to the appropriate segment line.

Table 3. Shutdown Register Format (Address (Hex) = 0xXC)

MODE	ADDRESS CODE (HEX)	REGISTER DATA							
		D7	D6	D5	D4	D3	D2	D1	D0
Shutdown Mode	0xXC	X	X	X	X	X	X	X	0
Normal Operation	0xXC	X	X	X	X	X	X	X	1

Table 4. Decode-Mode Register Examples (Address (Hex) = 0xX9)

DECODE MODE	REGISTER DATA								HEX CODE
	D7	D6	D5	D4	D3	D2	D1	D0	
No decode for digits 7-0	0	0	0	0	0	0	0	0	0x00
Code B decode for digit 0 No decode for digits 7-1	0	0	0	0	0	0	0	1	0x01
Code B decode for digits 3-0 No decode for digits 7-4	0	0	0	0	1	1	1	1	0x0F
Code B decode for digits 7-0	1	1	1	1	1	1	1	1	0xFF

MAX7219/MAX7221
Serially Interfaced, 8-Digit LED Display Drivers

Table 5. Code B Font

7-SEGMENT CHARACTER	REGISTER DATA						ON SEGMENTS = 1							
	D7*	D6–D4	D3	D2	D1	D0	DP*	A	B	C	D	E	F	G
0		X	0	0	0	0	0	1	1	1	1	1	1	0
1		X	0	0	0	1	0	0	1	1	0	0	0	0
2		X	0	0	1	0	0	1	1	0	1	1	0	1
3		X	0	0	1	1	0	1	1	1	1	0	0	1
4		X	0	1	0	0	0	0	1	1	0	0	1	1
5		X	0	1	0	1	0	1	0	1	1	0	1	1
6		X	0	1	1	0	0	1	0	1	1	1	1	1
7		X	0	1	1	1	0	1	1	1	0	0	0	0
8		X	1	0	0	0	0	1	1	1	1	1	1	1
9		X	1	0	0	1	0	1	1	1	1	0	1	1
—		X	1	0	1	0	0	0	0	0	0	0	0	1
E		X	1	0	1	1	0	1	0	0	1	1	1	1
H		X	1	1	0	0	0	0	1	1	0	1	1	1
L		X	1	1	0	1	0	0	0	0	1	1	1	0
P		X	1	1	1	0	0	1	1	0	0	1	1	1
blank		X	1	1	1	1	0	0	0	0	0	0	0	0

*The decimal point is set by bit D7 = 1.

Table 6. No-Decode Mode Data Bits and Corresponding Segment Lines

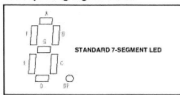

STANDARD 7-SEGMENT LED

	REGISTER DATA							
	D7	D6	D5	D4	D3	D2	D1	D0
Corresponding Segment Line	DP	A	B	C	D	E	F	G

Intensity Control and Interdigit Blanking

The MAX7219/MAX7221 allow display brightness to be controlled with an external resistor (R$_{SET}$) connected between V+ and ISET. The peak current sourced from the segment drivers is nominally 100 times the current entering ISET. This resistor can either be fixed or variable to allow brightness adjustment from the front panel. Its minimum value should be 9.53kΩ, which typically sets the segment current at 40mA. Display brightness can also be controlled digitally by using the intensity register.

Digital control of display brightness is provided by an internal pulse-width modulator, which is controlled by the lower nibble of the intensity register. The modulator scales the average segment current in 16 steps from a maximum of 31/32 down to 1/32 of the peak current set by R$_{SET}$ (15/16 to 1/16 on MAX7221). Table 7 lists the intensity register format. The minimum interdigit blanking time is set to 1/32 of a cycle.

Table 7. Intensity Register Format (Address (Hex) = 0xXA)

DUTY CYCLE		D7	D6	D5	D4	D3	D2	D1	D0	HEX CODE
MAX7219	MAX7221									
1/32 (min on)	1/16 (min on)	X	X	X	X	0	0	0	0	0xX0
3/32	2/16	X	X	X	X	0	0	0	1	0xX1
5/32	3/16	X	X	X	X	0	0	1	0	0xX2
7/32	4/16	X	X	X	X	0	0	1	1	0xX3
9/32	5/16	X	X	X	X	0	1	0	0	0xX4
11/32	6/16	X	X	X	X	0	1	0	1	0xX5
13/32	7/16	X	X	X	X	0	1	1	0	0xX6
15/32	8/16	X	X	X	X	0	1	1	1	0xX7
17/32	9/16	X	X	X	X	1	0	0	0	0xX8
19/32	10/16	X	X	X	X	1	0	0	1	0xX9
21/32	11/16	X	X	X	X	1	0	1	0	0xXA
23/32	12/16	X	X	X	X	1	0	1	1	0xXB
25/32	13/16	X	X	X	X	1	1	0	0	0xXC
27/32	14/16	X	X	X	X	1	1	0	1	0xXD
29/32	15/16	X	X	X	X	1	1	1	0	0xXE
31/32	15/16 (max on)	X	X	X	X	1	1	1	1	0xXF

Table 8. Scan-Limit Register Format (Address (Hex) = 0xXB)

SCAN LIMIT	REGISTER DATA								HEX CODE
	D7	D6	D5	D4	D3	D2	D1	D0	
Display digit 0 only*	X	X	X	X	X	0	0	0	0xX0
Display digits 0 & 1*	X	X	X	X	X	0	0	1	0xX1
Display digits 0 1 2*	X	X	X	X	X	0	1	0	0xX2
Display digits 0 1 2 3	X	X	X	X	X	0	1	1	0xX3
Display digits 0 1 2 3 4	X	X	X	X	X	1	0	0	0xX4
Display digits 0 1 2 3 4 5	X	X	X	X	X	1	0	1	0xX5
Display digits 0 1 2 3 4 5 6	X	X	X	X	X	1	1	0	0xX6
Display digits 0 1 2 3 4 5 6 7	X	X	X	X	X	1	1	1	0xX7

*See Scan-Limit Register section for application.

Scan-Limit Register

The scan-limit register sets how many digits are displayed, from 1 to 8. They are displayed in a multiplexed manner with a typical display scan rate of 800Hz with 8 digits displayed. If fewer digits are displayed, the scan rate is $8f_{OSC}/N$, where N is the number of digits scanned. Since the number of scanned digits affects the display brightness, the scan-limit register should not be used to blank portions of the display (such as leading zero suppression). Table 8 lists the scan-limit register format.

Serially Interfaced, 8-Digit LED Display Drivers

If the scan-limit register is set for three digits or less, individual digit drivers will dissipate excessive amounts of power. Consequently, the value of the R_{SET} resistor must be adjusted according to the number of digits displayed, to limit individual digit driver power dissipation. Table 9 lists the number of digits displayed and the corresponding maximum recommended segment current when the digit drivers are used.

Display-Test Register

The display-test register operates in two modes: normal and display test. Display-test mode turns all LEDs on by overriding, but not altering, all controls and digit registers (including the shutdown register). In display-test mode, 8 digits are scanned and the duty cycle is 31/32 (15/16 for MAX7221). Table 10 lists the display-test register format.

Table 9. Maximum Segment Current for 1-, 2-, or 3-Digit Displays

NUMBER OF DIGITS DISPLAYED	MAXIMUM SEGMENT CURRENT (mA)
1	10
2	20
3	30

Table 10. Display-Test Register Format (Address (Hex) = 0xXF)

MODE	REGISTER DATA							
	D7	D6	D5	D4	D3	D2	D1	D0
Normal Operation	X	X	X	X	X	X	X	0
Display Test Mode	X	X	X	X	X	X	X	1

Note: The MAX7219/MAX7221 remain in display-test mode (all LEDs on) until the display-test register is reconfigured for normal operation.

No-Op Register

The no-op register is used when cascading MAX7219s or MAX7221s. Connect all devices' LOAD/CS inputs together and connect DOUT to DIN on adjacent devices. DOUT is a CMOS logic-level output that easily drives DIN of successively cascaded parts. (Refer to the Serial Addressing Modes section for detailed information on serial input/output timing.) For example, if four MAX7219s are cascaded, then to write to the fourth chip, sent the desired 16-bit word, followed by three no-op codes (hex 0xXX0X, see Table 2). When LOAD/CS goes high, data is latched in all devices. The first three chips receive no-op commands, and the fourth receives the intended data.

Applications Information

Supply Bypassing and Wiring

To minimize power-supply ripple due to the peak digit driver currents, connect a 10μF electrolytic and a 0.1μF ceramic capacitor between V+ and GND as close to the device as possible. The MAX7219/MAX7221 should be placed in close proximity to the LED display, and connections should be kept as short as possible to minimize the effects of wiring inductance and electromagnetic interference. Also, both GND pins must be connected to ground.

Selecting R_{SET} Resistor and Using External Drivers

The current per segment is approximately 100 times the current in ISET. To select R_{SET}, see Table 11. The MAX7219/MAX7221's maximum recommended segment current is 40mA. For segment current levels above these levels, external digit drivers will be needed. In this application, the MAX7219/MAX7221 serve only as controllers for other high-current drivers or transistors. Therefore, to conserve power, use $R_{SET} = 47k\Omega$ when using external current sources as segment drivers.

The example in Figure 2 uses the MAX7219/MAX7221's segment drivers, a MAX394 single-pole double-throw analog switch, and external transistors to drive 2.3" AND2307SLC common-cathode displays. The 5.6V zener diode has been added in series with the decimal point LED because the decimal point LED forward voltage is typically 4.2V. For all other segments the LED forward voltage is typically 8V. Since external transistors are used to sink current (DIG 0 and DIG 1 are used as logic switches), peak segment currents of 45mA are allowed even though only two digits are displayed. In applications where the MAX7219/MAX7221's digit drivers are used to sink current and fewer than four digits are displayed, Table 9 specifies the maximum allowable segment current. R_{SET} must be selected accordingly (Table 11).

Refer to the Continuous Power Dissipation section of the Absolute Maximum Ratings to calculate acceptable limits for ambient temperature, segment current, and the LED forward-voltage drop.

Table 11. R$_{SET}$ vs. Segment Current and LED Forward Voltage

I$_{SEG}$ (mA)	V$_{LED}$ (V)				
	1.5	2.0	2.5	3.0	3.5
40	12.2	11.8	11.0	10.6	9.69
30	17.8	17.1	15.8	15.0	14.0
20	29.8	28.0	25.9	24.5	22.6
10	66.7	63.7	59.3	55.4	51.2

Table 12. Package Thermal Resistance Data

PACKAGE	THERMAL RESISTANCE (θ$_{JA}$)
24 Narrow DIP	+75°C/W
24 Wide SO	+85°C/W
24 CERDIP	+80°C/W
Maximum Junction Temperature (T$_J$) = +150°C	
Maximum Ambient Temperature (T$_A$) = +85°C	

Computing Power Dissipation

The upper limit for power dissipation (PD) for the MAX7219/MAX7221 is determined from the following equation:

$$PD = (V+ \times 8mA) + (V+ - V_{LED})(DUTY \times I_{SEG} \times N)$$

where:

V+ = supply voltage

DUTY = duty cycle set by intensity register

N = number of segments driven (worst case is 8)

V$_{LED}$ = LED forward voltage

I$_{SEG}$ = segment current set by R$_{SET}$

Dissipation example:

I$_{SEG}$ = 40mA, N = 8, DUTY = 31/32, V$_{LED}$ = 1.8V at 40mA, V+ = 5.25V

$$PD = (5.25V \times 8mA) + (5.25V - 1.8V)(31/32 \times 40mA \times 8) = 1.11W$$

Thus, for a CERDIP package (θ$_{JA}$ = +80°C/W from Table 12), the maximum allowed ambient temperature T$_A$ is given by:

$$T_{J(MAX)} = T_A + PD \times θ_{JA}$$

150°C = T$_A$ +1.11W x 80°C/W

where T$_A$ = +61.2°C.

The T$_A$ limits for PDIP and SO packages in the dissipation example above are +66.7°C and +55.6°C, respectively.

Cascading Drivers

The example in Figure 3 drives 16 digits using a 3-wire µP interface. If the number of digits is not a multiple of 8, set both drivers' scan limits registers to the same number so one display will not appear brighter than the other. For example, if 12 digits are need, use 6 digits per display with both scan-limit registers set for 6 digits so that both displays have a 1/6 duty cycle per digit. If 11 digits are needed, set both scan-limit registers for 6 digits and leave one digit driver unconnected. If one display for 6 digits and the other for 5 digits, the second display will appear brighter because its duty cycle per digit will be 1/5 while the first display's will be 1/6. Refer to the *No-Op Register* section for additional information.

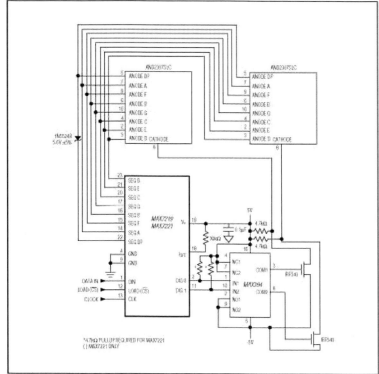

Figure 2. MAX7219/MAX7221 Driving 2.3in Displays

MAX7219/MAX7221
Serially Interfaced, 8-Digit LED Display Drivers

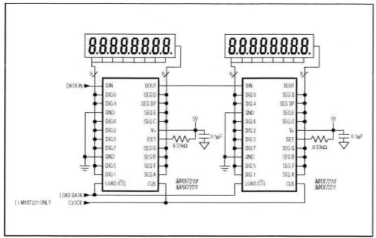

Figure 3. Cascading MAX7219/MAX7221s to Drive 16 Seven-Segment LED Digits

MAX7219/MAX7221
Serially Interfaced, 8-Digit LED Display Drivers

_____Chip Topography

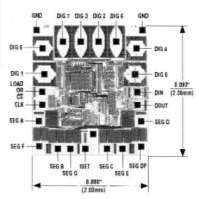

TRANSISTOR COUNT: 5267
SUBSTRATE CONNECTED TO GND

Maxim Integrated

Package Information

(The package drawing(s) in this data sheet may not reflect the most current specifications. For the latest package outline information go to www.maxim-ic.com/packages.)

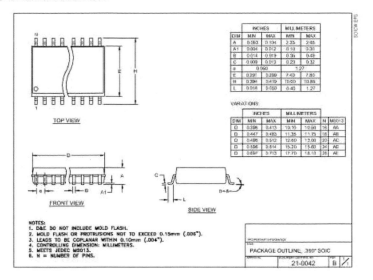

	INCHES		MILLIMETERS	
DIM	MIN	MAX	MIN	MAX
A	0.093	0.104	2.35	2.65
A1	0.004	0.012	0.10	0.30
B	0.014	0.019	0.35	0.49
C	0.009	0.013	0.23	0.32
e	0.050		1.27	
E	0.291	0.299	7.40	7.60
H	0.394	0.419	10.00	10.65
L	0.016	0.050	0.40	1.27

VARIATIONS

	INCHES		MILLIMETERS			
DIM	MIN	MAX	MIN	MAX	N	MS013
D	0.398	0.413	10.10	10.50	16	AA
D	0.447	0.463	11.35	11.75	18	AB
D	0.496	0.512	12.60	13.00	20	AC
D	0.598	0.614	15.20	15.60	24	AD
D	0.697	0.713	17.70	18.10	28	AE

TOP VIEW

FRONT VIEW

SIDE VIEW

NOTES:
1. D&E DO NOT INCLUDE MOLD FLASH.
2. MOLD FLASH OR PROTRUSIONS NOT TO EXCEED 0.15mm (.006").
3. LEADS TO BE COPLANAR WITHIN 0.10mm (.004").
4. CONTROLLING DIMENSION: MILLIMETERS.
5. MEETS JEDEC MS013.
6. N = NUMBER OF PINS.

PROPRIETARY INFORMATION

TITLE
PACKAGE OUTLINE, .300" SOIC

APPROVAL	DOCUMENT CONTROL NO.	REV	
	21-0042	B	1/1

Package Information (continued)

(The package drawing(s) in this data sheet may not reflect the most current specifications. For the latest package outline information go to www.maxim-ic.com/packages.)

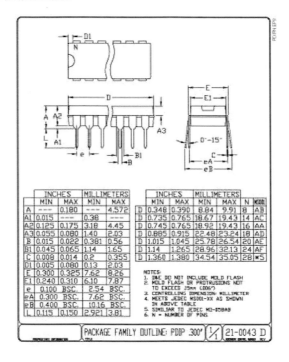

	INCHES		MILLIMETERS	
	MIN	MAX	MIN	MAX
A	---	0.180	---	4.572
A1	0.015	---	0.38	---
A2	0.125	0.175	3.18	4.45
A3	0.055	0.080	1.40	2.03
B	0.015	0.022	0.381	0.56
B1	0.045	0.065	1.14	1.65
C	0.008	0.014	0.2	0.355
D1	0.005	0.080	0.13	2.03
E	0.300	0.325	7.62	8.26
E1	0.240	0.310	6.10	7.87
e	0.100 BSC.		2.54 BSC.	
eA	0.300 BSC.		7.62 BSC.	
eB	0.400 BSC.		10.16 BSC.	
L	0.115	0.150	2.921	3.81

	INCHES		MILLIMETERS			
	MIN	MAX	MIN	MAX	N	MSDO
D	0.348	0.390	8.84	9.91	8	AB
D	0.735	0.765	18.67	19.43	14	AC
D	0.745	0.765	18.92	19.43	16	AA
D	0.885	0.915	22.48	23.24	18	AD
D	1.015	1.045	25.78	26.54	20	AE
D	1.14	1.265	28.96	32.13	24	AF
D	1.360	1.380	34.54	35.05	28	*5

NOTES:
1. D&E DO NOT INCLUDE MOLD FLASH.
2. MOLD FLASH OR PROTRUSIONS NOT TO EXCEED .15mm (.006")
3. CONTROLLING DIMENSION: MILLIMETER
4. MEETS JEDEC MS001-XX AS SHOWN IN ABOVE TABLE
5. SIMILAR TO JEDEC MO-05BAB
6. N = NUMBER OF PINS

PACKAGE FAMILY OUTLINE: PDIP .300"	1/1	21-0043 D

MAX7219/MAX7221
Serially Interfaced, 8-Digit LED Display Drivers

資料來源：http://pdfserv.maximintegrated.com/en/ds/MAX7219-MAX7221.pdf

RGB 控制板線路圖

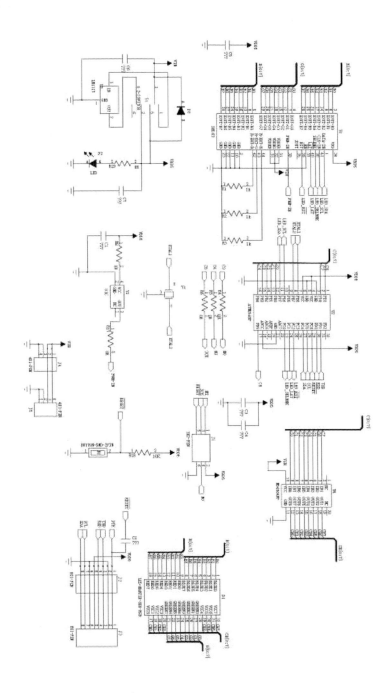

DM163 資料手冊

　　本書使用的 DM163 資料手冊，乃是點晶科技股份有限公司出產的 DM163 Full-Color LED Display ，在其公司網站有相關的資料手冊，讀者可以到 http://www.siti.com.tw/product/product6_en.html 下載其 DM163 Full-Color LED Display 函式庫資料手冊，特感謝點晶科技股份有限公司（ 參考網址： http://www.siti.com.tw/) 。

DM163

Version : A.004
Issue Date : 2005/8/19
File Name : SP-DM163-A.004.doc
Total Pages: 21

8x3-CHANNEL CONSTANT CURRENT LED DRIVER

新竹市科學園區展業一路 9 號 7 樓之 1
SILICON TOUCH TECHNOLOGY INC.
9-7F-1, Prosperity Road I, Science Based Industrial Park,
Hsin-Chu, Taiwan 300, R.O.C.
Tel : 886-3-5649656   Fax : 886-3-5649626

DM163

8x3-CHANNEL CONSTANT CURRENT LED DRIVERS

General Description

The DM163 is a LED driver that comprises shift registers, data latches, 8x3-channel constant current circuitry with current value set by 3 external resistors, and 64 x 256 gray level PWM (Pulse Width Modulation) function unit. Each channel provides a maximum current of 60 mA. The grayscale data are separated into BANK0 and BANK1 respectively, selected by SELBK pin. BANK0 is 6-bits grayscale data and the BNAK1 is 8-bits grayscale data. Depending on the system requirement, both PWM banks could be utilized jointly to achieve maximum 8+6 bit grayscale performance. Alternatively, users can choose either 64-graylevel bank or 256-graylevel bank for dot correction, and the remaining bank as image data.

DM163 could also be constructed as a PWM controller for LED drivers. When VDDH is connected to VDD, each of the 24 output channels outputs can act as an inverse digital signal for controlling the LED driver.

Features

- 24 Output Channels
- 8 + 6-bits PWM grayscale Control
- Constant Current Output: 5mA to 60mA
- LED Power Supply Voltage up to 17V
- VDD=3V to 5.5V
- Varied Output Current Level Set By 3 External Resistors
- Serial Shift-In Architecture for Grayscale Data

Block Diagram

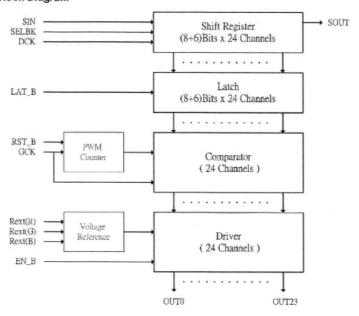

Figure 1. Functional Schematic of Whole Chip

The schematic of DM163 comprises of several fundamental units as shown in Figure 1. The grayscale data are input onto the DM163 by the **SIN** pin and transferred according to the synchronous clock **DCK**. Meanwhile, in order to separate the data into two groups, **SELBK** is designed as a switch control pin. When a sequence of data is already transferred onto the chip, the **LAT_B**='H' is set to convey it into the comparator unit. Compared with the counter signals, the grayscale data will determine the PWM control signal to display varied luminance at driver output. The **Rext** resistors are able to set diverse output current levels. The detailed schematic of each channel is shown as Figure 2.

Block Diagram

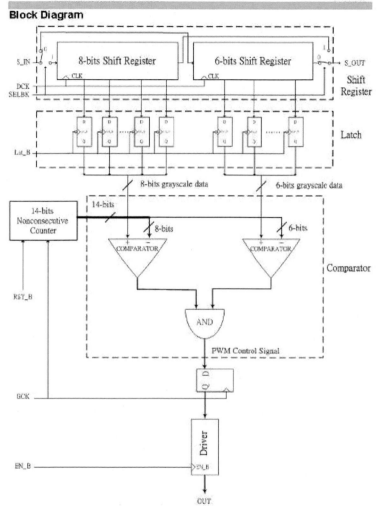

Figure 2. The Detailed Schematic of Each Channel

Pin Connection (Top view)

QFP44

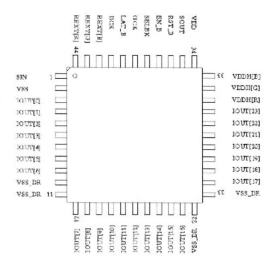

Pin No.	NAME	Pin No.	NAME	Pin No.	NAME	Pin No.	NAME
1	SIN	12	IOUT[7]	23	VSS_DR	34	VDD
2	VSS	13	IOUT[8]	24	IOUT[17]	35	SOUT
3	IOUT[0]	14	IOUT[9]	25	IOUT[18]	36	RST_B
4	IOUT[1]	15	IOUT[10]	26	IOUT[19]	37	EN_B
5	IOUT[2]	16	IOUT[11]	27	IOUT[20]	38	SELBK
6	IOUT[3]	17	IOUT[12]	28	IOUT[21]	39	GCK
7	IOUT[4]	18	IOUT[13]	29	IOUT[22]	40	LAT_B
8	IOUT[5]	19	IOUT[14]	30	IOUT[23]	41	DCK
9	IOUT[6]	20	IOUT[15]	31	VDDH[R]	42	REXT[B]
10	VSS_DR	21	IOUT[16]	32	VDDH[G]	43	REXT[G]
11	VSS_DR	22	VSS_DR	33	VDDH[B]	44	REXT[R]

QFN40

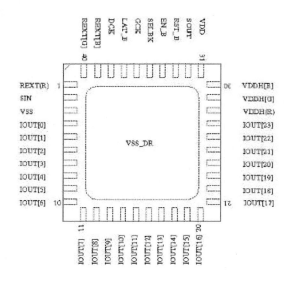

Pin No.	NAME	Pin No.	NAME	Pin No.	NAME	Pin No.	NAME
1	REXT[R]	11	IOUT[7]	21	IOUT[17]	31	VDD
2	SIN	12	IOUT[8]	22	IOUT[18]	32	SOUT
3	VSS	13	IOUT[9]	23	IOUT[19]	33	RST_B
4	IOUT[0]	14	IOUT[10]	24	IOUT[20]	34	EN_B
5	IOUT[1]	15	IOUT[11]	25	IOUT[21]	35	SELBK
6	IOUT[2]	16	IOUT[12]	26	IOUT[22]	36	GCK
7	IOUT[3]	17	IOUT[13]	27	IOUT[23]	37	LAT_B
8	IOUT[4]	18	IOUT[14]	28	VDDH[R]	38	DCK
9	IOUT[5]	19	IOUT[15]	29	VDDH[G]	39	REXT[B]
10	IOUT[6]	20	IOUT[16]	30	VDDH[B]	40	REXT[G]

Pin Description

PIN NAME	FUNCTION	QFP pin number	QFN pin number
VDDH (R)	Output protection pins.	31	28
VDDH (G)	They could be connected independently	32	29
VDDH (B)	or to LED supplies (VLED).	33	30
VDD	Power supply terminal.	34	31
VSS	Ground terminal.	2	3
VSS_DR	Driver ground	10, 11, 22, 23	Thermal pad
SIN	Serial input for grayscale data.	1	2
SOUT	Serial output for grayscale data.	35	32
DCK	Synchronous clock input for serial data transfer. The input data of SIN is transferred at rising edges of DCK.	41	38
SELBK	If SELBK is H, shift-in date would be stored in the 8-bit BANK 1. If SELBK is L, shift-in date would be stored in the 6-bit BANK 0.	38	35
LAT_B	When LAT_B converts from H to L, grayscale data in both shift register banks are latched.	40	37
GCK	Clock input for PWM operation.	39	36
$R_{EXT}(R)$ $R_{EXT}(G)$ $R_{EXT}(B)$	External resistor connected between R_{EXT} and GND for driver current setting. $R_{EXT}(R)$ controls outputs OUT0, 3, 6, 9, 12, 15, 18, 21. $R_{EXT}(G)$ controls outputs OUT1, 4, 7, 10, 13, 16, 19, 22. $R_{EXT}(B)$ controls outputs OUT2, 5, 8, 11, 14, 17, 20, 23.	44 43 42	1 40 39
IOUT0~23	LED driver outputs.	3, 4, 5, 6, 7, 8, 9, 12, 13, 14, 15,16, 17, 18, 19, 20,21, 24, 25, 26, 27,28, 29, 30	4, 5, 6, 7, 8, 9, 10, 11, 12, 13, 14, 15, 16, 17, 18, 19, 20, 21, 22, 23, 24, 25, 26, 27
EN_B	Input terminal of output enable. All outputs are OFF when EN_B is H.	37	34
RST_B	The IC is initialized when RST_B low. There is an internal pull-up on this pin. This pin couldn't be floating. Before using the IC, it must be reset first. If each channel is assigned to drive multiple LEDs, IC should be reset before each LED data latch to prevent from flashing.	36	33

Maximum Ratings (Ta=25°C, Tj(max) = 140°C)

CHARACTERISTIC	SYMBOL	RATING	UNIT
Supply Voltage	VDD	-0.3 ~ 7.0	V
Input Voltage	VIN	-0.3 ~ VDD+0.3	V
Output Current	IOUT	60	mA
Output Voltage	VOUT	-0.3 ~ 17	V
DCK Frequency	FDCK	20	MHz
GCK Frequency	FGCK	20	MHz
GND Terminal Current	IGND	1440	mA
Power Dissipation	PD	1.36 (QFP44); 3.63 (QFN40) (Ta=25°C)	W
Thermal Resistance	Rth(j-a)	84.42 (QFP44); 31.67 (QFN40)	°C/W
Operating Temperature	Top	-40 ~ 85	°C
Storage Temperature	Tstg	-55 ~ 150	°C

Recommended Operating Condition

DC Characteristics (Ta = 25°C)

CHARACTERISTIC	SYMBOL	CONDITION	MIN.	TYP.	MAX.	UNIT
Supply Voltage	VDD	——	3		5.5	V
Output Voltage	VOUT	——	——		17	V
Output Current	Io	OUTn	5	——	60	
	IOH	SERIAL-OUT	——	——	2	mA
	IOL	SERIAL-OUT	——	——	-2	
Input Voltage	VIH	——	0.8 VDD	——	VDD+0.2	V
	VIL		-0.2	——	0.2 VDD	

AC Characteristics (VDD = 5.0 V, Ta = 25°C)

CHARACTERISTIC	SYMBOL	CONDITION	MIN.	TYP.	MAX.	UNIT
DCK Frequency	FDCK	Cascade operation	——	——	20	MHz
DCK pulse duration	t_{wh} / t_{wl}	High or low level	15	——	——	ns
DCK rise/fall time	t_r / t_f		——	——	20	ns
GCK Frequency	FGCK	——	1	——	20	MHz
GCK pulse duration	t_{wh} / t_{wl}	High or low level	15	——	——	ns
GCK rise/fall time	t_r / t_f	——	——	——	20	ns
Set-up Time for SIN	tsetup(D)	Before DCK rising edge	2	——	——	ns
Hold Time for SIN	thold(D)	After DCK rising edge	3	——	——	ns
Set-up Time for DCK	tsetup(L)	Before LAT_B falling edge	3	——	——	ns
LAT_B Pulse Width	tw LAT	——	5	——	——	ns
Set-up Time for LAT_B	Tsetup(G)	Before GCK rising edge	13	——	——	ns
Set-up Time for SELBK	Tsetup(S)	Before DCK rising edge	5	——	——	ns
Hold Time for SELBK	Thold(S)	After DCK rising edge	1	——	——	ns

8x3-CHANNEL CONSTANT CURRENT LED DRIVERS   **Version: A.004**   Page 7

AC Characteristics (V$_{DD}$ = 3.3 V, Ta = 25°C)

CHARACTERISTIC	SYMBOL	CONDITION	MIN.	TYP.	MAX.	UNIT
DCK Frequency	F$_{DCK}$	Cascade operation	---	---	20	MHz
DCK pulse duration	t$_{wh}$ / t$_{wl}$	High or low level	15	---	---	ns
DCK rise/fall time	t$_r$ / t$_f$		---	---	20	ns
GCK Frequency	F$_{GCK}$	Cascade operation	1	---	20	MHz
GCK pulse duration	t$_{wh}$ / t$_{wl}$	High or low level	15	---	---	ns
GCK rise/fall time	t$_r$ / t$_f$	---	---	---	20	ns
RST_B pulse duration	twrst_b	Low level	100	---	---	ns
Set-up Time for SIN	tsetup(D)	Before DCK rising edge	2	---	---	ns
Hold Time for SIN	thold(D)	After DCK rising edge	5	---	---	ns
Set-up Time for DCK	tsetup(L)	Before LAT_B falling edge	5	---	---	ns
LAT_B Pulse Width	tw LAT		7	---	---	ns
Set-up Time for LAT_B	Tsetup(G)	Before GCK rising edge	23	---	---	ns
Set-up Time for SELBK	Tsetup(S)	Before DCK rising edge	9	---	---	ns
Hold Time for SELBK	Thold(S)	After DCK rising edge	1	---	---	ns

Electrical Characteristics (V$_{DD}$ = 5.0 V, Ta = 25°C unless otherwise noted)

CHARACTERISTIC	SYMBOL	CONDITION	MIN.	TYP.	MAX.	UNIT
Input Voltage "H" Level	V$_{IH}$	---	0.8 V$_{DD}$	---	V$_{DD}$	V
Input Voltage "L" Level	V$_{IL}$	---	GND	---	0.2 V$_{DD}$	
Output Leakage Current	I$_{leak}$	V$_{OH}$ = 17 V	---	---	±0.1	uA
Output Voltage (SOUT)	V$_{OL}$	I$_{OL}$ = 2 mA	---	---	0.2	V
	V$_{OH}$	I$_{OH}$ = -2 mA	4.8	---	---	
Output Current (Channel-Channel)	I$_{OL1}$	V$_{OUT}$ = 1.0V R$_{EXT}$ = 2.6kΩ	---	±3	±5	%
Output Current (Chip-Chip)	I$_{OL3}$	V$_{OUT}$ = 1.0V R$_{EXT}$ = 2.6kΩ	---	±4	±10	%
Supply Voltage Regulation	% / V$_{DD}$	R$_{EXT}$ = 3kΩ	---	---	2	% / V
Supply Current[1]	I$_{DD, analog}$	VDD=5V, R$_{EXT}$ = 1kΩ	---	42.2	43.4	mA
	I$_{DD, digital}$	VDD=5V, Cload=2pF, DCK=GCK=1MHz	---	1	1.5	

---

[1] I$_{LED}$ excluded.

Switching Characteristics (V_{DD} = 3.3V, Ta = 25°C)

CHARACTERISTIC	SYMBOL	CONDITION	MIN.	TYP.	MAX.	UNIT
SOUT Rise time	t_{or}		—	4	5	ns
SOUT Fall time	t_{of}	VIH=VDD VIL=GND REXT=3KΩ CL=13pF	—	4	5	ns
SOUT Propagation delay (L to H)	t_{pLH}		—	24	30	ns
SOUT Propagation delay (H to L)	t_{pHL}		—	20	25	ns
IOUT Rise time	t_{or}		—	15	18	ns
IOUT Fall time	t_{of}	VIH=VDD VIL=GND REXT=3KΩ VLED=3.3V RL=120Ω CL=33pF	—	20	25	ns
IOUT Propagation delay After GCK or EN_B (L to H / OFF to ON)	t_{pLH}		—	35	37	ns
IOUT Propagation delay After GCK or EN_B (H to L / ON to OFF)	t_{pHL}		—	30	35	ns

Switching Characteristics (V_{DD} = 5.0V, Ta = 25°C)

CHARACTERISTIC	SYMBOL	CONDITION	MIN.	TYP.	MAX.	UNIT
SOUT Rise time	t_{or}		—	4	5	ns
SOUT Fall time	t_{of}	VIH=VDD VIL=GND REXT=3KΩ CL=13pF	—	4	6	ns
SOUT Propagation delay (L to H)	t_{pLH}		—	19	25	ns
SOUT Propagation delay (H to L)	t_{pHL}		—	17	23	ns
IOUT Rise time	t_{or}		—	4	6	ns
IOUT Fall time	t_{of}	VIH=VDD VIL=GND REXT=3KΩ VLED=5.0V RL=120Ω CL=33pF	—	15	18	ns
IOUT Propagation delay After GCK or EN_B (L to H / OFF to ON)	t_{pLH}		—	26	30	ns
IOUT Propagation delay After GCK or EN_B (H to L / ON to OFF)	t_{pHL}		—	20	25	ns

Input Capacitance (Ta = 25°C)

INPUT NODE	SYMBOL	CONDITION	MIN.	TYP.	MAX.	UNIT
SIN	C_{SIN}	—	—	3	—	pF
DCK	C_{DCK}	—	—	3	—	pF
GCK	C_{GCK}	—	—	3	—	pF
LAT_B	C_{LAT_B}	—	—	3	—	pF
EN_B	C_{EN_B}	—	—	3	—	pF
RST_B	C_{RST_B}	—	—	3	—	pF
SELBK	C_{SELBK}	—	—	3	—	pF

Parameter Measurement

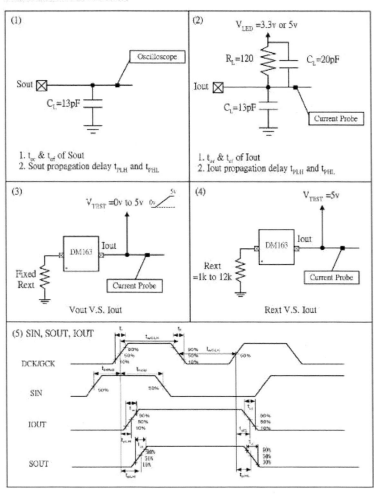

Serial Shift-In Luminance Data (Shift Register Architecture)

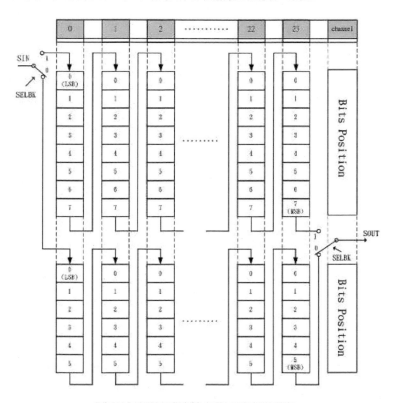

Figure 3. Serial Shift-In Luminance Data Structure

This serial shift (shift register) architecture follows a FIFO (first-in first-out) formate. The MSB (Most Significant Bit), both 8th bit and 6th bit at the 23rd channel, is the first data bit that shift into the driver. And the LSB (Least Significant Bit) data, the 1st bit at the 1st channel, is the last bit in the data sequence. Furthermore, the SELBK control signal is set to determine in which bank the data are placed.

Timing Diagram

Timing diagram

Assumption: 64-graylevel(6-bit) as correction terms, 256-graylevel(8-bit) as image data, N pcs, DM163 connected in series

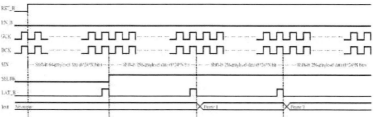

Figure 4. Timing diagram when 6bits are correction terms and 8bits are image terms

When 6 bits are correction terms and 8 bits are image terms (as shown in Fig 4), users must set the controller signals according to below sequences:

(1) Set SELBK=L (Bank 0) and begin shift in 6 bits correction data

(2) Set LAT_B=H to update the correction data after all correction data are in place

(3) Set SELBK=H (Bank 1) and begin shift in 8 bits image data

(4) Set LAT_B=H to update image data after 8 bit image are all in place. DM163 will utilize the 8 bits image data to determine the grayscale of each channel

(5) Repeat steps (3) and (4)

Timing diagram

Assumption: 64-graylevel(6-bit) and 256-graylevel(8-bit) are both image data, N pcs, DM163 connected in series

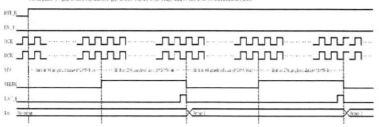

Figure 5. Timing diagram when both 6bits and 8bits are used as image terms

When both 6 bits and 8 bits bank are used for images terms (As shown in Fig. 5), users should set the controller signal in accordance to the following:

(1) Set SELBK=L (Bank 0) and begin shift in 6 bits correction data
(2) Set SELBK=H (Bank 1) and begin shift in 8 bits image data
(3) Set LAT_B=H to update image data after both 8 bit and 6 bit image data are all inplace.
(4) Repeat steps (1) to (3)

Timing Diagram

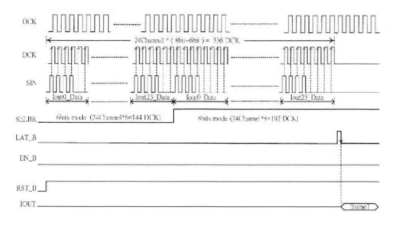

Figure 6. Detailed timing diagram of data transference

Figure 6 shows the detailed timing diagram of data transference. The synchronous clock DCK is designed to trigger at the positive edge. And the LAT_B triggers at the negative edge. To completely fill up both 6 bit and 8 bit shift register, a total of 336 DCK count is required (144 DCK for 6bits mode and 192 DCK for 8bits mode). Example depicted in figure 6 shows 6'b001111 data at 6bits bank and 8'b00001111 at 8bits bank respectively. Therefore, the average output current is (15/256) x (15/64) x Iout.

Formula I (out, avg)= (BANK 1/256) x (BANK 0/64) x Iout, provides a useful way to calculate the input data and the output current. Iout is the reference current value shown in figure 12. Users could utilize the formula Iout =47*Vrext / Rext to get an approximate value of Iout.

Particular Phenomenon

DM163 incorporates a different PWM counter, as described in Figure 2, hence its output waveform demonstrate a very different characteristics compare to conventional PWM counter.

(1) Nonconsecutive counter

The non-consecutive PWM counter incorporate by DM163 demonstrated a waveform pattern similar to Figure 7. Its waveform is spread-out into each PWM cycle, resulting lots of intermediate pulses during each PWM cycle. In Fig 7, if all the intermediate pulses are added up, it would equal to 50% luminance which is the same as the conventional method. By spreading out the PWM pulses, this approach can help prevent LED from flickering in lower grayscale situation.

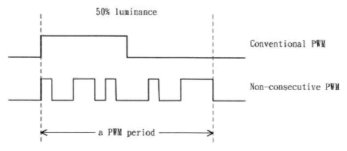

Figure 7. An Example of Nonconsecutive PWM Signal

(2) 8+6 bits Comparator

The comparator illustrated in Fig 2 is another one of the unique designs in DM163. The comparator's output will be "H" only when value at "+" is larger then the value at "-" (in other word, comparator will be "L" when value in "+" equals to value in "-" or value in "+" is less than value in "-"). Only when both 8 bit and 6 bit comparator are "H" will there be current in the output channel.

Due to this unique comparator design, DM163 exhibit a very distinct output characters in two certain scenario. In the first case, DM163 output will always be "OFF" when either one of the 8 bit or 6 bit bank is filled with 0. In 2nd scenario, when all bit value at both 8 bit and 6 bit bank are loaded "H", DM163 output will exhibit its highest luminance value (but not 100% luminance value). Due to the nature of comparators design, PWM control signal will be zero in the condition of 8bits counter=8'bFF or 6bits counter=6'b3F. Consequently, the PWM control signal will be 0 for $2^{8}+2^{6}+1$ GCK rather than always high.

Application Diagram

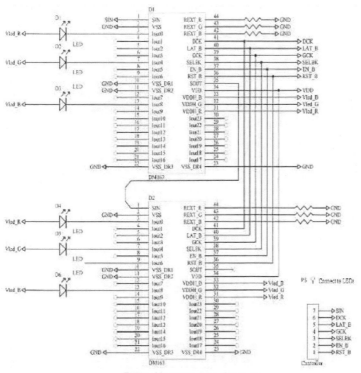

Figure 10. Application Diagram

Note:

1. The RST_B should be connected to controller to initialize the IC.
2. VDDH_R/G/B should be connected to Vled_R/G/B respectively.   The Vled_R/G/B are power supply of Red/Green/Blue LEDs.
3. VSS_DR is the ground pin of LEDs.   And it could be connected to VSS.

Application Diagram (Cont.)

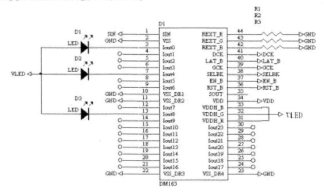

Figure 11. Application Diagram of anode-common LED

Driver Output Current (V_{DD} = 3.3V and 5.0V, Ta = 25°C)

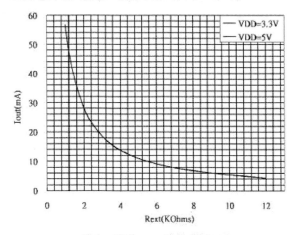

Figure 12. R_{EXT} vs. Output Current

- 434 -

Driver Output Current (Cont.)

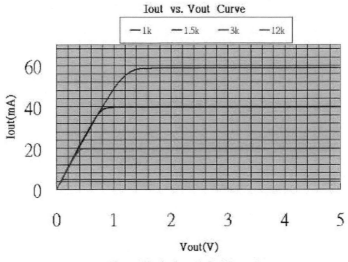

Figure 13. Vout vs. Output Current

The curve shown in Fig. 12 is the average result of a large number of samples. Due to chip-to-chip variation in Vrext, users may observe a different Iout-Vout curve than above. However, the curves of VDD=5v and VDD=3.3v should be close to each other when the same chip is tested because DM163 utilizes a negative feedback circuit to keep the average voltage of Vrext pins close to constant, regardless of the VDD. Therefore, the Iout-to-Rext curve should not be seriously influenced by VDD variation.

The Fig. 13 illustrates the relation between Vout and Iout. Iout is the constant value when Vout exceeds the voltage of turning point. In other words, Iout is independent of the fluctuation of Vout if IC is biased in this condition.

Package Outline Dimension

QFP44

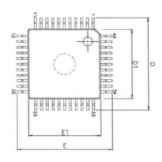

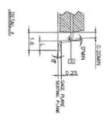
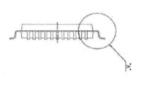

UNIT : mm

SYMBOLS	MIN	NOM	MAX
A	–	–	2.7
A1	0.25	0.30	0.35
A2	1.9	2.0	2.2
b	–	0.3 (TYP)	–
D	13.00	13.20	13.40
D1	9.9	10.00	10.10
E	13.00	13.20	13.40
E1	9.9	10.00	10.10
e	–	0.80 (TYP)	–
b	0.73	0.88	0.93
θ	0	–	7
c	0.1	0.15	0.2

NOTES:

1. JEDEC OUTLINE: MO-108 AA-1

2. DATUM PLANE B IS LOCATED AT THE BOTTOM OF THE MOLD PARTING LINE COINCIDENT WITH WHERE THE LEAD EXITS THE BODY.

3. DIMENSIONS D1 AND E1 DO NOT INCLUDE MOLD PROTRUSION. ALLOWABLE PROTRUSION IS 0.25 mm PER SIDE. DIMENSIONS D1 AND E1 DO INCLUDE MOLD MISMATCH AND ARE DETERMINED AT DATUM PLANE B.

4. DIMENSION b DOES NOT INCLUDE DAMBAR PROTRUSION.

QFN40

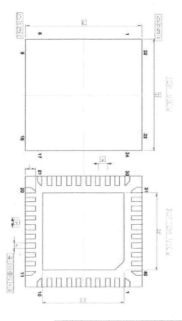

SYMBOL	DIMENSION (MM)			DIMENSION (MIL)		
	MIN	NOM	MAX	MIN	NOM	MAX
A						
A1	0	0.02	0.05	0	0.8	2.0
A3		0.25 REF			9.84 REF	
b	0.18	0.23	0.30	7.1	9.1	11.8
D		6.00 BSC			236.2 BSC	
D2	4.50	4.60	4.80	68.9	181.1	189
E		6.00 BSC			236.2 BSC	
E2	3.70	4.25	4.25	145.7	167.3	167.3
e		0.50 BSC			19.7 BSC	
L	0.35	0.40	1.50	13.8	15.8	19.7
Y		0.30			3.9	

The products listed herein are designed for ordinary electronic applications, such as electrical appliances, audio-visual equipment, communications devices and so on. Hence, it is advisable that the devices should not be used in medical instruments, surgical implants, aerospace machinery, nuclear power control systems, disaster/crime-prevention equipment and the like. Misusing those products may directly or indirectly endanger human life, or cause injury and property loss.

Silicon Touch Technology, Inc. will not take any responsibilities regarding the misusage of the products mentioned above. Anyone who purchases any products described herein with the above-mentioned intention or with such misused applications should accept full responsibility and indemnify. Silicon Touch Technology, Inc. and its distributors and all their officers and employees shall defend jointly and severally against any and all claims and litigation and all damages, cost and expenses associated with such intention and manipulation.

TFT 顯示模組

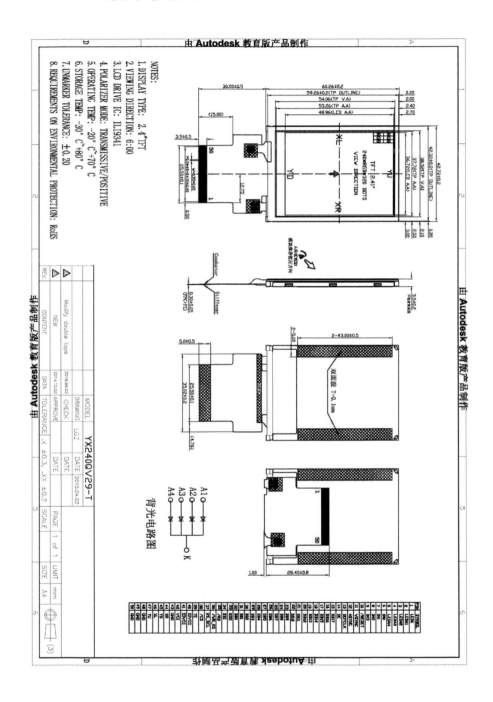

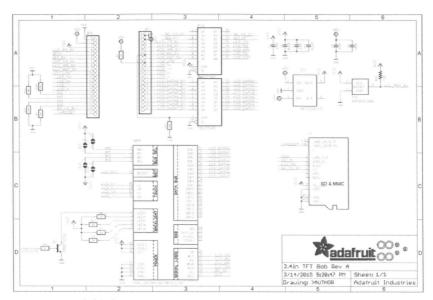

資料來源：Adafruit 2.4" Color TFT Touchscreen Breakout，網址：
https://learn.adafruit.com/adafruit-2-4-color-tft-touchscreen-breakout/overview?view=all#assembly

Nokia 5110 LCD

DATA SHEET

PCD8544
48 × 84 pixels matrix LCD controller/driver

Product specification
File under Integrated Circuits, IC17

1999 Apr 12

Philips
Semiconductors

PHILIPS

48 × 84 pixels matrix LCD controller/driver   PCD8544

CONTENTS

48 × 84 pixels matrix LCD controller/driver

PCD8544

1 FEATURES

- Single chip LCD controller/driver
- 48 row, 84 column outputs
- Display data RAM 48 × 84 bits
- On-chip:
  - Generation of LCD supply voltage (external supply also possible)
  - Generation of intermediate LCD bias voltages
  - Oscillator requires no external components (external clock also possible).
- External $\overline{\text{RES}}$ (reset) input pin
- Serial interface maximum 4.0 Mbits/s
- CMOS compatible inputs
- Mux rate: 48
- Logic supply voltage range V_{DD} to V_{SS}: 2.7 to 3.3 V
- Display supply voltage range V_{LCD} to V_{SS}
  - 6.0 to 8.5 V with LCD voltage internally generated (voltage generator enabled)
  - 6.0 to 9.0 V with LCD voltage externally supplied (voltage generator switched-off).
- Low power consumption, suitable for battery operated systems
- Temperature compensation of V_{LCD}
- Temperature range: −25 to +70 °C.

2 GENERAL DESCRIPTION

The PCD8544 is a low power CMOS LCD controller/driver, designed to drive a graphic display of 48 rows and 84 columns. All necessary functions for the display are provided in a single chip, including on-chip generation of LCD supply and bias voltages, resulting in a minimum of external components and low power consumption.

The PCD8544 interfaces to microcontrollers through a serial bus interface.

The PCD8544 is manufactured in n-well CMOS technology.

3 APPLICATIONS

- Telecommunications equipment.

4 ORDERING INFORMATION

TYPE NUMBER	PACKAGE			
	NAME	DESCRIPTION	VERSION
PCD8544U	—	chip with bumps in tray; 168 bonding pads + 4 dummy pads	—

48 × 84 pixels matrix LCD controller/driver

PCD8544

5 BLOCK DIAGRAM

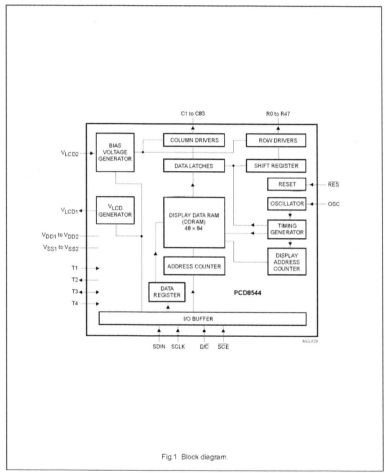

Fig.1 Block diagram.

48 × 84 pixels matrix LCD controller/driver

PCD8544

6 PINNING

SYMBOL	DESCRIPTION
R0 to R47	LCD row driver outputs
C0 to C83	LCD column driver outputs
V_{SS1}, V_{SS2}	ground
V_{DD1}, V_{DD2}	supply voltage
V_{LCD1}, V_{LCD2}	LCD supply voltage
T1	test 1 input
T2	test 2 output
T3	test 3 input/output
T4	test 4 input
SDIN	serial data input
SCLK	serial clock input
D/$\overline{C}$	data/command
$\overline{SCE}$	chip enable
OSC	oscillator
$\overline{RES}$	external reset input
dummy1, 2, 3, 4	not connected

Note

1. For further details, see Fig.18 and Table 7.

6.1 Pin functions

6.1.1 R0 TO R47 ROW DRIVER OUTPUTS

These pads output the row signals.

6.1.2 C0 TO C83 COLUMN DRIVER OUTPUTS

These pads output the column signals.

6.1.3 V_{SS1}, V_{SS2}: NEGATIVE POWER SUPPLY RAILS

Supply rails V_{SS1} and V_{SS2} must be connected together.

6.1.4 V_{DD1}, V_{DD2}: POSITIVE POWER SUPPLY RAILS

Supply rails V_{DD1} and V_{DD2} must be connected together.

6.1.5 V_{LCD1}, V_{LCD2}: LCD POWER SUPPLY

Positive power supply for the liquid crystal display. Supply rails V_{LCD1} and V_{LCD2} must be connected together.

6.1.6 T1, T2, T3 AND T4: TEST PADS

T1, T3 and T4 must be connected to V_{SS}, T2 is to be left open. Not accessible to user.

6.1.7 SDIN: SERIAL DATA LINE

Input for the data line.

6.1.8 SCLK: SERIAL CLOCK LINE

Input for the clock signal: 0.0 to 4.0 Mbits/s.

6.1.9 D/$\overline{C}$: MODE SELECT

Input to select either command/address or data input.

6.1.10 $\overline{SCE}$: CHIP ENABLE

The enable pin allows data to be clocked in. The signal is active LOW.

6.1.11 OSC: OSCILLATOR

When the on-chip oscillator is used, this input must be connected to V_{DD}. An external clock signal, if used, is connected to this input. If the oscillator and external clock are both inhibited by connecting the OSC pin to V_{SS}, the display is not clocked and may be left in a DC state. To avoid this, the chip should always be put into Power-down mode before stopping the clock.

6.1.12 $\overline{RES}$: RESET

This signal will reset the device and must be applied to properly initialize the chip. The signal is active LOW.

48 × 84 pixels matrix LCD controller/driver

PCD8544

7 FUNCTIONAL DESCRIPTION

7.1 Oscillator

The on-chip oscillator provides the clock signal for the display system. No external components are required and the OSC input must be connected to V_{DD}. An external clock signal, if used, is connected to this input.

7.2 Address Counter (AC)

The address counter assigns addresses to the display data RAM for writing. The X-address X_6 to X_0 and the Y-address Y_2 to Y_0 are set separately. After a write operation, the address counter is automatically incremented by 1, according to the V flag.

7.3 Display Data RAM (DDRAM)

The DDRAM is a 48 × 84 bit static RAM which stores the display data. The RAM is divided into six banks of 84 bytes (6 × 8 × 84 bits). During RAM access, data is transferred to the RAM through the serial interface. There is a direct correspondence between the X-address and the column output number.

7.4 Timing generator

The timing generator produces the various signals required to drive the internal circuits. Internal chip operation is not affected by operations on the data buses.

7.5 Display address counter

The display is generated by continuously shifting rows of RAM data to the dot matrix LCD through the column outputs. The display status (all dots on/off and normal/inverse video) is set by bits E and D in the 'display control' command.

7.6 LCD row and column drivers

The PCD8544 contains 48 row and 84 column drivers, which connect the appropriate LCD bias voltages in sequence to the display in accordance with the data to be displayed. Figure 2 shows typical waveforms. Unused outputs should be left unconnected.

48 × 84 pixels matrix LCD controller/driver

PCD8544

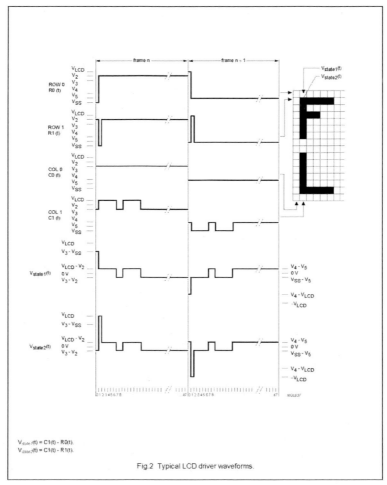

$V_{state1}(t) = C1(t) - R0(t)$.
$V_{state2}(t) = C1(t) - R1(t)$.

Fig.2 Typical LCD driver waveforms.

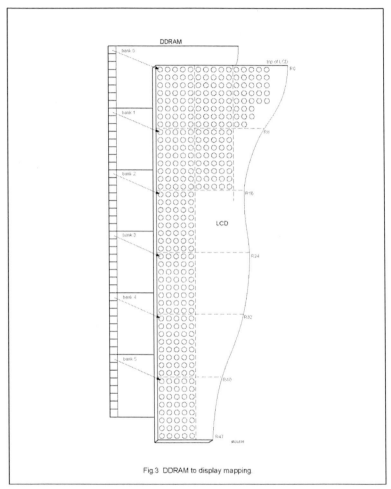

Fig.3 DDRAM to display mapping.

48 × 84 pixels matrix LCD controller/driver

PCD8544

7.7 Addressing

Data is downloaded in bytes into the 48 by 84 bits RAM data display matrix of PCD8544, as indicated in Figs. 3, 4, 5 and 6. The columns are addressed by the address pointer. The address ranges are: X 0 to 83 (1010011), Y 0 to 5 (101). Addresses outside these ranges are not allowed. In the vertical addressing mode (V = 1), the Y address increments after each byte (see Fig.5). After the last Y address (Y = 5), Y wraps around to 0 and X increments to address the next column. In the horizontal addressing mode (V = 0), the X address increments after each byte (see Fig.6). After the last X address (X = 83), X wraps around to 0 and Y increments to address the next row. After the very last address (X = 83 and Y = 5), the address pointers wrap around to address (X = 0 and Y = 0).

7.7.1 DATA STRUCTURE

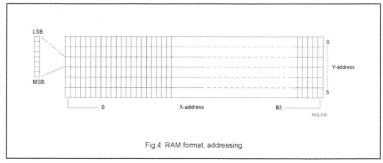

Fig.4 RAM format, addressing.

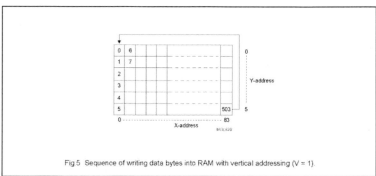

Fig.5 Sequence of writing data bytes into RAM with vertical addressing (V = 1).

48 × 84 pixels matrix LCD controller/driver

PCD8544

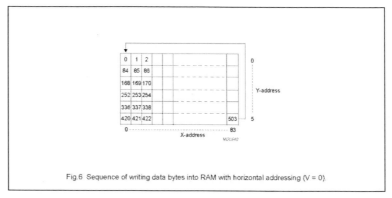

Fig.6 Sequence of writing data bytes into RAM with horizontal addressing (V = 0).

7.8 Temperature compensation

Due to the temperature dependency of the liquid crystals' viscosity, the LCD controlling voltage V_{LCD} must be increased at lower temperatures to maintain optimum contrast. Figure 7 shows V_{LCD} for high multiplex rates. In the PCD8544, the temperature coefficient of V_{LCD} can be selected from four values (see Table 2) by setting bits TC_1 and TC_0.

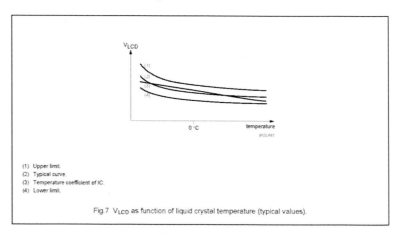

(1) Upper limit.
(2) Typical curve.
(3) Temperature coefficient of IC.
(4) Lower limit.

Fig.7 V_{LCD} as function of liquid crystal temperature (typical values).

48 × 84 pixels matrix LCD controller/driver

PCD8544

8 INSTRUCTIONS

The instruction format is divided into two modes: If D/$\overline{\text{C}}$ (mode select) is set LOW, the current byte is interpreted as command byte (see Table 1). Figure 8 shows an example of a serial data stream for initializing the chip. If D/$\overline{\text{C}}$ is set HIGH, the following bytes are stored in the display data RAM. After every data byte, the address counter is incremented automatically.

The level of the D/$\overline{\text{C}}$ signal is read during the last bit of data byte.

Each instruction can be sent in any order to the PCD8544. The MSB of a byte is transmitted first. Figure 9 shows one possible command stream, used to set up the LCD driver.

The serial interface is initialized when $\overline{\text{SCE}}$ is HIGH. In this state, SCLK clock pulses have no effect and no power is consumed by the serial interface. A negative edge on $\overline{\text{SCE}}$ enables the serial interface and indicates the start of a data transmission.

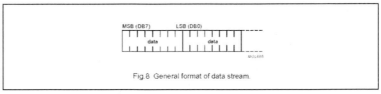

Fig.8 General format of data stream.

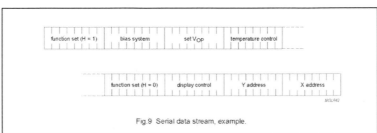

Fig.9 Serial data stream, example.

Figures 10 and 11 show the serial bus protocol.

- When $\overline{\text{SCE}}$ is HIGH, SCLK clock signals are ignored; during the HIGH time of $\overline{\text{SCE}}$, the serial interface is initialized (see Fig.12).
- SDIN is sampled at the positive edge of SCLK.
- D/$\overline{\text{C}}$ indicates whether the byte is a command (D/$\overline{\text{C}}$ = 0) or RAM data (D/$\overline{\text{C}}$ = 1); it is read with the eighth SCLK pulse.

- If $\overline{\text{SCE}}$ stays LOW after the last bit of a command/data byte, the serial interface expects bit 7 of the next byte at the next positive edge of SCLK (see Fig.12).
- A reset pulse with $\overline{\text{RES}}$ interrupts the transmission. No data is written into the RAM. The registers are cleared. If $\overline{\text{SCE}}$ is LOW after the positive edge of $\overline{\text{RES}}$, the serial interface is ready to receive bit 7 of a command/data byte (see Fig.13).

48 × 84 pixels matrix LCD controller/driver

PCD8544

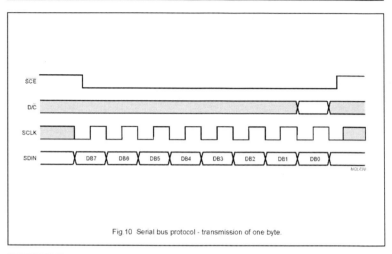

Fig 10 Serial bus protocol - transmission of one byte.

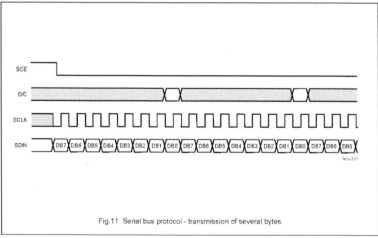

Fig.11 Serial bus protocol - transmission of several bytes.

- 452 -

48 × 84 pixels matrix LCD controller/driver

PCD8544

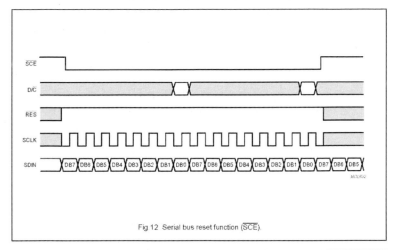

Fig 12 Serial bus reset function (SCE).

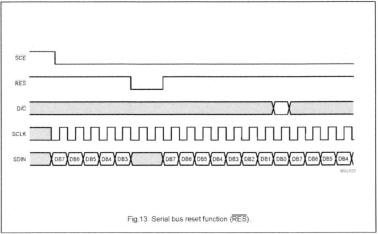

Fig.13 Serial bus reset function (RES).

48 × 84 pixels matrix LCD controller/driver

PCD8544

Table 1 Instruction set

INSTRUCTION	D/C̄	COMMAND BYTE								DESCRIPTION
		DB7	DB6	DB5	DB4	DB3	DB2	DB1	DB0	
(H = 0 or 1)										
NOP	0	0	0	0	0	0	0	0	0	no operation
Function set	0	0	0	1	0	0	PD	V	H	power down control; entry mode; extended instruction set control (H)
Write data	1	D_7	D_6	D_5	D_4	D_3	D_2	D_1	D_0	writes data to display RAM
(H = 0)										
Reserved	0	0	0	0	0	0	1	X	X	do not use
Display control	0	0	0	0	0	1	D	0	E	sets display configuration
Reserved	0	0	0	0	1	X	X	X	X	do not use
Set Y address of RAM	0	0	1	0	0	0	Y_2	Y_1	Y_0	sets Y-address of RAM; $0 \leq Y \leq 5$
Set X address of RAM	0	1	X_6	X_5	X_4	X_3	X_2	X_1	X_0	sets X-address part of RAM; $0 \leq X \leq 83$
(H = 1)										
Reserved	0	0	0	0	0	0	0	0	1	do not use
	0	0	0	0	0	0	0	1	X	do not use
Temperature control	0	0	0	0	0	0	1	TC_1	TC_0	set Temperature Coefficient (TC_x)
Reserved	0	0	0	0	0	1	X	X	X	do not use
Bias system	0	0	0	0	1	0	BS_2	BS_1	BS_0	set Bias System (BS_x)
Reserved	0	0	1	X	X	X	X	X	X	do not use
Set V_{OP}	0	1	V_{OP6}	V_{OP5}	V_{OP4}	V_{OP3}	V_{OP2}	V_{OP1}	V_{OP0}	write V_{OP} to register

Table 2 Explanations of symbols in Table 1

BIT	0	1
PD	chip is active	chip is in Power-down mode
V	horizontal addressing	vertical addressing
H	use basic instruction set	use extended instruction set
D and E		
00	display blank	
10	normal mode	
01	all display segments on	
11	inverse video mode	
TC_1 and TC_0		
00	V_{LCD} temperature coefficient 0	
01	V_{LCD} temperature coefficient 1	
10	V_{LCD} temperature coefficient 2	
11	V_{LCD} temperature coefficient 3	

8.1 Initialization

Immediately following power-on, the contents of all internal registers and of the RAM are undefined. **A RES pulse must be applied**. Attention should be paid to the possibility that the **device may be damaged** if not properly reset.

All internal registers are reset by applying an external $\overline{RES}$ pulse (active LOW) at pad 31, within the specified time. However, the RAM contents are still undefined. The state after reset is described in Section 8.2.

The $\overline{RES}$ input must be $\leq 0.3 V_{DD}$ when V_{DD} reaches V_{DDmin} (or higher) within a maximum time of 100 ms after V_{DD} goes HIGH (see Fig.16).

8.2 Reset function

After reset, the LCD driver has the following state:

- Power-down mode (bit PD = 1)
- Horizontal addressing (bit V = 0) normal instruction set (bit H = 0)
- Display blank (bit E = D = 0)
- Address counter X_6 to $X_0 = 0$; Y_2 to $Y_0 = 0$
- Temperature control mode (TC_1 $TC_0 = 0$)
- Bias system (BS_2 to $BS_0 = 0$)
- V_{LCD} is equal to 0, the HV generator is switched off (V_{OP6} to $V_{OP0} = 0$)
- After power-on, the RAM contents are undefined.

8.3 Function set

8.3.1 BIT PD

- All LCD outputs at V_{SS} (display off)
- Bias generator and V_{LCD} generator off, V_{LCD} can be disconnected
- Oscillator off (external clock possible)
- Serial bus, command, etc. function
- Before entering Power-down mode, the RAM needs to be filled with '0's to ensure the specified current consumption.

8.3.2 BIT V

When V = 0, the horizontal addressing is selected. The data is written into the DDRAM as shown in Fig.6. When V = 1, the vertical addressing is selected. The data is written into the DDRAM, as shown in Fig.5.

8.3.3 BIT H

When H = 0 the commands 'display control', 'set Y address' and 'set X address' can be performed; when H = 1, the others can be executed. The 'write data' and 'function set' commands can be executed in both cases.

8.4 Display control

8.4.1 BITS D AND E

Bits D and E select the display mode (see Table 2).

8.5 Set Y address of RAM

Y_n defines the Y vector addressing of the display RAM.

Table 3 Y vector addressing

Y_2	Y_1	Y_0	BANK
0	0	0	0
0	0	1	1
0	1	0	2
0	1	1	3
1	0	0	4
1	0	1	5

8.6 Set X address of RAM

The X address points to the columns. The range of X is 0 to 83 (53H).

8.7 Temperature control

The temperature coefficient of V_{LCD} is selected by bits TC_1 and TC_0.

8.8 Bias value

The bias voltage levels are set in the ratio of R - R - nR - R - R, giving a 1/(n + 4) bias system. Different multiplex rates require different factors n (see Table 4). This is programmed by BS_2 to BS_0. For Mux 1 : 48, the optimum bias value n, resulting in 1/8 bias, is given by:

$$n = \sqrt{48} - 3 = 3.928 = 4 \qquad (1)$$

48 × 84 pixels matrix LCD controller/driver

PCD8544

Table 4 Programming the required bias system

BS$_2$	BS$_1$	BS$_0$	n	RECOMMENDED MUX RATE
0	0	0	7	1 : 100
0	0	1	6	1 : 80
0	1	0	5	1 : 65/1 : 65
0	1	1	4	1 : 48
1	0	0	3	1 : 40/1 : 34
1	0	1	2	1 : 24
1	1	0	1	1 : 18/1 : 16
1	1	1	0	1 : 10/1 : 9/1 : 8

Table 5 LCD bias voltage

SYMBOL	BIAS VOLTAGES	BIAS VOLTAGE FOR $\frac{1}{8}$ BIAS
V1	V_{LCD}	V_{LCD}
V2	$(n + 3)/(n + 4)$	$\frac{7}{8} \times V_{LCD}$
V3	$(n + 2)/(n + 4)$	$\frac{6}{8} \times V_{LCD}$
V4	$2/(n + 4)$	$\frac{2}{8} \times V_{LCD}$
V5	$1/(n + 4)$	$\frac{1}{8} \times V_{LCD}$
V6	V_{SS}	V_{SS}

8.9 Set V_{OP} value

The operation voltage V_{LCD} can be set by software. The values are dependent on the liquid crystal selected. $V_{LCD} = a + (V_{OP6}$ to $V_{OP0}) \times b$ [V]. In the PCD8544, a = 3.06 and b = 0.06 giving a program range of 3.00 to 10.68 at room temperature.

Note that the charge pump is turned off if V_{OP6} to V_{OP0} is set to zero.

For Mux 1 : 48, the optimum operation voltage of the liquid can be calculated as:

$$V_{LCD} = \frac{1 + \sqrt{48}}{\sqrt{2 \cdot \left(1 - \frac{1}{\sqrt{48}}\right)}} \cdot V_{th} = 6.06 \cdot V_{th} \qquad (2)$$

where V_{th} is the threshold voltage of the liquid crystal material used.

Caution, as V_{OP} increases with lower temperatures, care must be taken not to set a V_{OP} that will exceed the maximum of 8.5 V when operating at −25 °C.

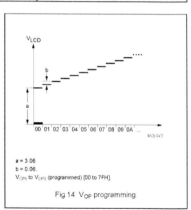

a = 3.06
b = 0.06.
V_{OP6} to V_{OP0} (programmed) [00 to 7FH].

Fig 14 V_{OP} programming

48 × 84 pixels matrix LCD controller/driver   PCD8544

9   LIMITING VALUES

In accordance with the Absolute Maximum Rating System (IEC 134); see notes 1 and 2.

SYMBOL	PARAMETER	CONDITIONS	MIN.	MAX.	UNIT
V_{DD}	supply voltage	note 3	-0.5	+7	V
V_{LCD}	supply voltage LCD	note 4	-0.5	+10	V
V_i	all input voltages		-0.5	$V_{DD} + 0.5$	V
I_{SS}	ground supply current		-50	+50	mA
I_I, I_O	DC input or output current		-10	+10	mA
P_{tot}	total power dissipation		-	300	mW
P_O	power dissipation per output		-	30	mW
T_{amb}	operating ambient temperature		-25	+70	°C
T_j	operating junction temperature		-65	+150	°C
T_{stg}	storage temperature		-65	+150	°C

Notes

1. Stresses above those listed under limiting values may cause permanent damage to the device.

2   Parameters are valid over operating temperature range unless otherwise specified. All voltages are with respect to V_{SS} unless otherwise noted.

3. With external LCD supply voltage externally supplied (voltage generator disabled). $V_{DDmax} = 5$ V if LCD supply voltage is internally generated (voltage generator enabled).

4. When setting V_{LCD} by software, take care not to set a V_{OP} that will exceed the maximum of 8.5 V when operating at -25 °C, see Caution in Section 8.9.

10 HANDLING

Inputs and outputs are protected against electrostatic discharge in normal handling. However, to be totally safe, it is desirable to take normal precautions appropriate to handling MOS devices (see *"Handling MOS devices"*).

48 × 84 pixels matrix LCD controller/driver

PCD8544

11 DC CHARACTERISTICS

V_{DD} = 2.7 to 3.3 V; V_{SS} = 0 V; V_{LCD} = 6.0 to 9.0 V; T_{amb} = −25 to +70 °C; unless otherwise specified.

SYMBOL	PARAMETER	CONDITIONS	MIN.	TYP.	MAX.	UNIT
V_{DD1}	supply voltage 1	LCD voltage externally supplied (voltage generator disabled)	2.7	–	3.3	V
V_{DD2}	supply voltage 2	LCD voltage internally generated (voltage generator enabled)	2.7	–	3.3	V
V_{LCD1}	LCD supply voltage	LCD voltage externally supplied (voltage generator disabled)	6.0	–	9.0	V
V_{LCD2}	LCD supply voltage	LCD voltage internally generated (voltage generator enabled); note 1	6.0	–	8.5	V
I_{DD1}	supply current 1 (normal mode) for internal V_{LCD}	V_{DD} = 2.85 V; V_{LCD} = 7.0 V; f_{SCLK} = 0; T_{amb} = 25 °C; display load = 10 μA; note 2	–	240	300	μA
I_{DD2}	supply current 2 (normal mode) for internal V_{LCD}	V_{DD} = 2.70 V; V_{LCD} = 7.0 V; f_{SCLK} = 0; T_{amb} = 25 °C; display load = 10 μA; note 2	–	–	320	μA
I_{DD3}	supply current 3 (Power-down mode)	with internal or external LCD supply voltage; note 3	–	1.5	–	μA
I_{DD4}	supply current external V_{LCD}	V_{DD} = 2.85 V; V_{LCD} = 9.0 V; f_{SCLK} = 0; notes 2 and 4	–	25	–	μA
I_{LCD}	supply current external V_{LCD}	V_{DD} = 2.7 V; V_{LCD} = 7.0 V; f_{SCLK} = 0; T = 25 °C; display load = 10 μA; notes 2 and 4	–	42	–	μA
Logic						
V_{IL}	LOW level input voltage		V_{SS}	–	$0.3V_{DD}$	V
V_{IH}	HIGH level input voltage		$0.7V_{DD}$	–	V_{DD}	V
I_L	leakage current	V_I = V_{DD} or V_{SS}	−1	–	+1	μA
Column and row outputs						
$R_{o(C)}$	column output resistance C0 to C83		–	12	20	kΩ
$R_{o(R)}$	row output resistance R0 to R47		–	12	20	kΩ
$V_{bias(tol)}$	bias voltage tolerance on C0 to C83 and R0 to R47		−100	0	+100	mV

48 × 84 pixels matrix LCD controller/driver

PCD8544

SYMBOL	PARAMETER	CONDITIONS	MIN.	TYP.	MAX.	UNIT
LCD supply voltage generator						
V_{LCD}	V_{LCD} tolerance internally generated	V_{DD} = 2.85 V; V_{LCD} = 7.0 V; f_{SCLK} = 0; display load = 10 μA; note 5	–	0	300	mV
TC0	V_{LCD} temperature coefficient 0	V_{DD} = 2.85 V; V_{LCD} = 7.0 V; f_{SCLK} = 0; display load = 10 μA	–	1	–	mV/K
TC1	V_{LCD} temperature coefficient 1	V_{DD} = 2.85 V; V_{LCD} = 7.0 V; f_{SCLK} = 0; display load = 10 μA	–	9	–	mV/K
TC2	V_{LCD} temperature coefficient 2	V_{DD} = 2.85 V; V_{LCD} = 7.0 V; f_{SCLK} = 0; display load = 10 μA	–	17	–	mV/K
TC3	V_{LCD} temperature coefficient 3	V_{DD} = 2.85 V; V_{LCD} = 7.0 V; f_{SCLK} = 0; display load = 10 μA	–	24	–	mV/K

Notes

1. The maximum possible V_{LCD} voltage that may be generated is dependent on voltage, temperature and (display) load.

2. Internal clock.

3. RAM contents equal '0'. During power-down, all static currents are switched off.

4. If external V_{LCD}, the display load current is not transmitted to I_{DD}.

5. Tolerance depends on the temperature (typically zero at 27 °C, maximum tolerance values are measured at the temperate range limit).

48 × 84 pixels matrix LCD controller/driver

PCD8544

12 AC CHARACTERISTICS

SYMBOL	PARAMETER	CONDITIONS	MIN.	TYP.	MAX.	UNIT
f_{OSC}	oscillator frequency		20	34	65	kHz
$f_{clk(ext)}$	external clock frequency		10	32	100	kHz
f_{frame}	frame frequency	f_{OSC} or $f_{clk(ext)}$ = 32 kHz; note 1	–	67	–	Hz
t_{VHRL}	V_{DD} to RES LOW	Fig.16	0[2]	–	30	ms
$t_{WL(RES)}$	RES LOW pulse width	Fig.16	100	–	–	ns
Serial bus timing characteristics						
f_{SCLK}	clock frequency	V_{DD} = 3.0 V ±10%	0	–	4.00	MHz
T_{cy}	clock cycle SCLK	All signal timing is based on	250	–	–	ns
t_{WH1}	SCLK pulse width HIGH	20% to 80% of V_{DD} and	100	–	–	ns
t_{WL1}	SCLK pulse width LOW	maximum rise and fall times of	100	–	–	ns
t_{su2}	SCE set-up time	10 ns	60	–	–	ns
t_{h2}	SCE hold time		100	–	–	ns
t_{WH2}	SCE min. HIGH time		100	–	–	ns
t_{h5}	SCE start hold time; note 3		100	–	–	ns
t_{su3}	D/C set-up time		100	–	–	ns
t_{h3}	D/C hold time		100	–	–	ns
t_{su4}	SDIN set-up time		100	–	–	ns
t_{h4}	SDIN hold time		100	–	–	ns

Notes

1. $T_{frame} = \dfrac{f_{clk(ext)}}{480}$

2. RES may be LOW before V_{DD} goes HIGH.

3. t_{h5} is the time from the previous SCLK positive edge (irrespective of the state of SCE) to the negative edge of SCE (see Fig.15).

48 × 84 pixels matrix LCD controller/driver

PCD8544

12.1 Serial interface

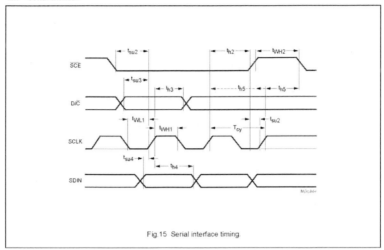

Fig.15 Serial interface timing.

12.2 Reset

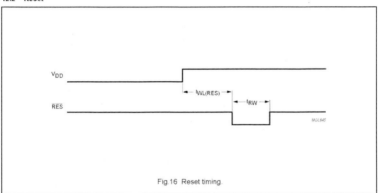

Fig.16 Reset timing.

48 × 84 pixels matrix LCD controller/driver

PCD8544

13 APPLICATION INFORMATION

Table 6 Programming example

STEP	SERIAL BUS BYTE									DISPLAY	OPERATION
	D/C̄	DB7	DB6	DB5	DB4	DB3	DB2	DB1	DB0		
1	start										SCE is going LOW
2	0	0	0	1	0	0	0	0	1		function set PD = 0 and V = 0, select extended instruction set (H = 1 mode)
3	0	1	0	0	1	0	0	0	0		set V_{OP}; V_{OP} is set to a +16 × b [V]
4	0	0	0	1	0	0	0	0	0		function set PD = 0 and V = 0, select normal instruction set (H = 0 mode)
5	0	0	0	0	0	1	1	0	0		display control set normal mode (D = 1 and E = 0)
6	1	0	0	0	1	1	1	1	1		data write Y and X are initialized to 0 by default, so they are not set here
7	1	0	0	0	0	0	1	0	1		data write
8	1	0	0	0	0	0	1	1	1		data write
9	1	0	0	0	0	0	0	0	0		data write
10	1	0	0	0	1	1	1	1	1		data write

48 × 84 pixels matrix LCD controller/driver   PCD8544

STEP	SERIAL BUS BYTE									DISPLAY	OPERATION
	D/C̄	DB7	DB6	DB5	DB4	DB3	DB2	DB1	DB0		
11	1	0	0	0	0	0	1	0	0		data write
12	1	0	0	0	1	1	1	1	1		data write
13	0	0	0	0	0	1	1	0	1		display control; set inverse video mode (D = 1 and E = 1)
14	0	1	0	0	0	0	0	0	0		set X address of RAM; set address to '0000000'
15	1	0	0	0	0	0	0	0	0		data write

The pinning is optimized for single plane wiring e.g. for chip-on-glass display modules. Display size: 48 × 84 pixels.

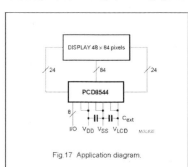

Fig.17 Application diagram.

The required minimum value for the external capacitors is: $C_{ext} = 1.0$ μF.

Higher capacitor values are recommended for ripple reduction.

14 BONDING PAD LOCATIONS

14.1 Bonding pad information (see Fig.18)

PARAMETER	SIZE
Pad pitch	min. 100 μm
Pad size, aluminium	80 × 100 μm
Bump dimensions	59 × 89 × 17.5 (±5) μm
Wafer thickness	max. 380 μm

14.2 Bonding pad location

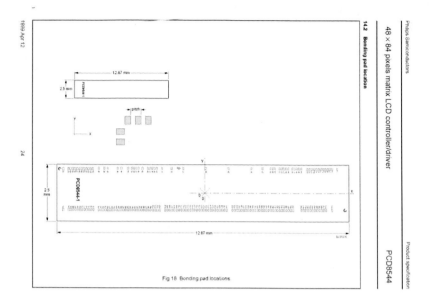

Fig.18 Bonding pad locations

- 464 -

48 × 84 pixels matrix LCD controller/driver

PCD8544

Table 7 Bonding pad locations (dimensions in µm).
All X/Y coordinates are referenced to the centre
of chip (see Fig.18)

PAD	PAD NAME	x	y
1	dummy1	+5932	+1060
2	R36	+5704	+1060
3	R37	+5604	+1060
4	R38	+5504	+1060
5	R39	+5404	+1060
6	R40	+5304	+1060
7	R41	+5204	+1060
8	R42	+5104	+1060
9	R43	+5004	+1060
10	R44	+4904	+1060
11	R45	+4804	+1060
12	R46	+4704	+1060
13	R47	+4604	+1060
14	V_{DD1}	+4330	+1085
15	V_{DD1}	+4230	+1085
16	V_{DD1}	+4130	+1085
17	V_{DD1}	+4030	+1085
18	V_{DD1}	+3930	+1085
19	V_{DD2}	+3750	+1085
20	V_{DD2}	+3650	+1085
21	V_{DD2}	+3550	+1085
22	V_{DD2}	+3450	+1085
23	V_{DD2}	+3350	+1085
24	V_{DD2}	+3250	+1085
25	V_{DD2}	+3150	+1085
26	V_{DD2}	+3050	+1085
27	SCLK	+2590	+1085
28	SDIN	+2090	+1085
29	D/$\overline{C}$	+1090	+1085
30	$\overline{SCE}$	+90	+1085
31	$\overline{RES}$	−910	+1085
32	OSC	−1410	+1085
33	T3	−1826	+1085
34	V_{SS2}	−2068	+1085
35	V_{SS2}	−2168	+1085
36	V_{SS2}	−2268	+1085
37	V_{SS2}	−2368	+1085
38	V_{SS2}	−2468	+1085

PAD	PAD NAME	x	y
39	T4	−2709	+1085
40	V_{SS1}	−2876	+1085
41	V_{SS1}	−2976	+1085
42	V_{SS1}	−3076	+1085
43	V_{SS1}	−3176	+1085
44	T1	−3337	+1085
45	V_{LCD2}	−3629	+1085
46	V_{LCD2}	−3789	+1085
47	V_{LCD1}	−4231	+1085
48	V_{LCD1}	−4391	+1085
49	T2	−4633	+1085
50	R23	−4894	+1060
51	R22	−4994	+1060
52	R21	−5094	+1060
53	R20	−5194	+1060
54	R19	−5294	+1060
55	R18	−5394	+1060
56	R17	−5494	+1060
57	R16	−5594	+1060
58	R15	−5694	+1060
59	R14	−5794	+1060
60	R13	−5894	+1060
61	R12	−5994	+1060
62	dummy2	−6222	+1060
63	dummy3	−6238	−738
64	R0	−5979	−738
65	R1	−5879	−738
66	R2	−5779	−738
67	R3	−5679	−738
68	R4	−5579	−738
69	R5	−5479	−738
70	R6	−5379	−738
71	R7	−5279	−738
72	R8	−5179	−738
73	R9	−5079	−738
74	R10	−4979	−738
75	R11	−4879	−738
76	C0	−4646	−746

48 × 84 pixels matrix LCD controller/driver

PCD8544

PAD	PAD NAME	x	y	PAD	PAD NAME	x	y
77	C1	-4546	-746	118	C42	-296	-746
78	C2	-4446	-746	119	C43	-196	-746
79	C3	-4346	-746	120	C44	-96	-746
80	C4	-4246	-746	121	C45	+4	-746
81	C5	-4146	-746	122	C46	+104	-746
82	C6	-4046	-746	123	C47	+204	-746
83	C7	-3946	-746	124	C48	+304	-746
84	C8	-3846	-746	125	C49	+404	-746
85	C9	-3746	-746	126	C50	+504	-746
86	C10	-3646	-746	127	C51	+604	-746
87	C11	-3546	-746	128	C52	+704	-746
88	C12	-3446	-746	139	C53	+804	-746
89	C13	-3346	-746	130	C54	+904	-746
90	C14	-3246	-746	131	C55	+1004	-746
91	C15	-3146	-746	132	C56	+1254	-746
92	C16	-3046	-746	133	C57	+1354	-746
93	C17	-2946	-746	134	C58	+1454	-746
94	C18	-2846	-746	135	C59	+1554	-746
95	C19	-2746	-746	136	C60	+1654	-746
96	C20	-2646	-746	137	C61	+1754	-746
97	C21	-2546	-746	138	C62	+1854	-746
98	C22	-2446	-746	139	C63	+1954	-746
99	C23	-2346	-746	140	C64	+2054	-746
100	C24	-2246	-746	141	C65	+2154	-746
101	C25	-2146	-746	142	C66	+2254	-746
102	C26	-2046	-746	143	C67	+2354	-746
103	C27	-1946	-746	144	C68	+2454	-746
104	C28	-1696	-746	145	C69	+2554	-746
105	C29	-1596	-746	146	C70	+2654	-746
106	C30	-1496	-746	147	C71	+2754	-746
107	C31	-1396	-746	148	C72	+2854	-746
108	C32	-1296	-746	149	C73	+2954	-746
109	C33	-1196	-746	150	C74	+3054	-746
110	C34	-1096	-746	151	C75	+3154	-746
111	C35	-996	-746	152	C76	+3254	-746
112	C36	-896	-746	153	C77	+3354	-746
113	C37	-796	-746	154	C78	+3454	-746
114	C38	-696	-746	155	C79	+3554	-746
115	C39	-596	-746	156	C80	+3654	-746
116	C40	-496	-746	157	C81	+3754	-746
117	C41	-396	-746	158	C82	+3854	-746

48 × 84 pixels matrix LCD controller/driver

PCD8544

PAD	PAD NAME	x	y
159	C83	+3954	−746
160	R35	+4328	−738
161	R34	+4428	−738
162	R33	+4528	−738
163	R32	+4628	−738
164	R31	+4728	−738
165	R30	+4828	−738
166	R29	+4928	−738
167	R28	+5028	−738
168	R27	+5128	−738
169	R26	+5228	−738
170	R25	+5328	−738
171	R24	+5428	−738
172	dummy4	+5694	−738

48 × 84 pixels matrix LCD controller/driver

PCD8544

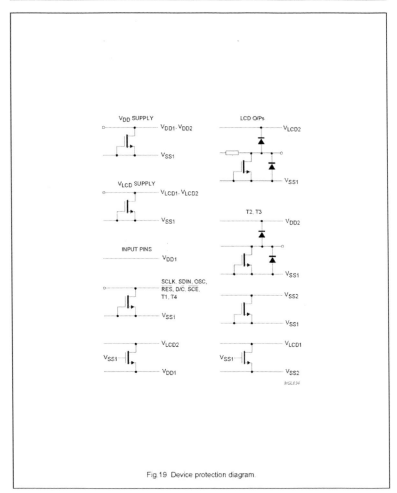

Fig.19 Device protection diagram.

48 × 84 pixels matrix LCD controller/driver

PCD8544

15 TRAY INFORMATION

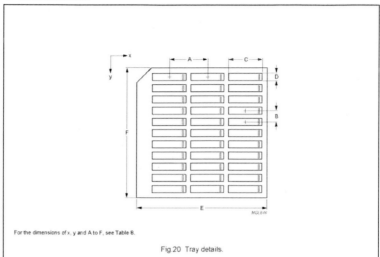

For the dimensions of x, y and A to F, see Table 8.

Fig.20 Tray details.

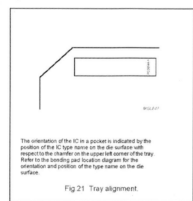

The orientation of the IC in a pocket is indicated by the position of the IC type name on the die surface with respect to the chamfer on the upper left corner of the tray. Refer to the bonding pad location diagram for the orientation and position of the type name on the die surface.

Fig 21 Tray alignment.

Table 8 Dimensions

DIM.	DESCRIPTION	VALUE
A	pocket pitch, in the x direction	14.82 mm
B	pocket pitch, in the y direction	4.39 mm
C	pocket width, in the x direction	13.27 mm
D	pocket width, in the y direction	2.8 mm
E	tray width, in the x direction	50.67 mm
F	tray width, in the y direction	50.67 mm
x	no. of pockets in the x direction	3
y	no. of pockets in the y direction	11

1999 Apr 12

29

48 × 84 pixels matrix LCD controller/driver

PCD8544

16 DEFINITIONS

Data sheet status	
Objective specification	This data sheet contains target or goal specifications for product development.
Preliminary specification	This data sheet contains preliminary data; supplementary data may be published later.
Product specification	This data sheet contains final product specifications.
Limiting values	
Limiting values given are in accordance with the Absolute Maximum Rating System (IEC 134). Stress above one or more of the limiting values may cause permanent damage to the device. These are stress ratings only and operation of the device at these or at any other conditions above those given in the Characteristics sections of the specification is not implied. Exposure to limiting values for extended periods may affect device reliability.	
Application information	
Where application information is given, it is advisory and does not form part of the specification.	

17 LIFE SUPPORT APPLICATIONS

These products are not designed for use in life support appliances, devices, or systems where malfunction of these products can reasonably be expected to result in personal injury. Philips customers using or selling these products for use in such applications do so at their own risk and agree to fully indemnify Philips for any damages resulting from such improper use or sale.

參考文獻

Anderson, R., & Cervo, D. (2013). *Pro Arduino*: Apress.

Arduino. (2013). Arduino official website. Retrieved from http://www.arduino.cc/

Atmel_Corporation. (2013). Atmel Corporation Website. Retrieved from http://www.atmel.com/

Banzi, M. (2009). *Getting Started with arduino*: Make.

Boxall, J. (2013). *Arduino Workshop: A Hands-on Introduction With 65 Projects*: No Starch Press.

Creative_Commons. (2013). Creative Commons. Retrieved from http://en.wikipedia.org/wiki/Creative_Commons

DFRobot. (2013). Arduino LCD KeyPad Shield Retrieved from http://www.dfrobot.com/wiki/index.php/Arduino_LCD_KeyPad_Shield_(SKU:_DFR0009)

Faludi, R. (2010). *Building wireless sensor networks: with ZigBee, XBee, arduino, and processing*: O'reilly.

Fritzing.org. (2013). Fritzing.org. Retrieved from http://fritzing.org/

Guangzhou_Tinsharp_Industrial_Corp._Ltd. (2013). TC1602A DataSheet. Retrieved from http://www.tinsharp.com/

Margolis, M. (2011). *Arduino cookbook*: O'Reilly Media.

Margolis, M. (2012). *Make an Arduino-controlled robot*: O'Reilly.

McRoberts, M. (2010). *Beginning Arduino*: Apress.

Minns, P. D. (2013). *C Programming For the PC the MAC and the Arduino Microcontroller System*: AuthorHouse.

Monk, S. (2010). 30 Arduino Projects for the Evil Genius, 2/e.

Monk, S. (2012). *Programming Arduino: Getting Started with Sketches*: McGraw-Hill.

Oxer, J., & Blemings, H. (2009). *Practical Arduino: cool projects for open source hardware*: Apress.

Reas, B. F. a. C. (2013). Processing. Retrieved from http://www.processing.org/

Reas, C., & Fry, B. (2007). *Processing: a programming handbook for visual designers and artists* (Vol. 6812): Mit Press.

Reas, C., & Fry, B. (2010). *Getting Started with Processing*: Make.

Warren, J.-D., Adams, J., & Molle, H. (2011). *Arduino for Robotics*: Springer.

Wilcher, D. (2012). *Learn electronics with Arduino*: Apress.

曹永忠. (2016a). 如何設計網路計時器：元件設計篇. *智慧家庭*. Retrieved from http://www.techbang.com/posts/43326-how-to-design-a-network-timer-component-design-review

曹永忠. (2016b). 物聯網系列：彩色顯示介紹(2.4~3.2"TFT 基本篇) *智慧家庭*. Retrieved from https://vmaker.tw/archives/10466

曹永忠. (2016c). 物聯網系列：彩色顯示介紹(2.4~3.2"TFT 進階篇). *智慧家庭*. Retrieved from https://vmaker.tw/archives/10827

曹永忠. (2016d). 物聯網系列：彩色顯示介紹（OLED LCD 篇）. *智慧家庭*. Retrieved from https://vmaker.tw/archives/8673

曹永忠. (2016e). 物聯網系列：單色圖形顯示介紹(NOKIA 5110 LCD 基本篇). *智慧家庭*. Retrieved from https://vmaker.tw/

曹永忠. (2016f). 物聯網系列：單色圖形顯示介紹(NOKIA 5110 LCD 開發篇). *智慧家庭*. Retrieved from https://vmaker.tw/

曹永忠. (2016g). 智慧家庭：如何安裝各類感測器的函式庫. *智慧家庭*. Retrieved from https://vmaker.tw/archives/3730

曹永忠. (2016h). 顯示技術：視覺暫留的應用- 手搖字幕機開發軟體篇. *顯示技術*. Retrieved from http://www.techbang.com/posts/42336-display-technologies-persistence-of-vision-hand-application-of-subtitle-software-article

曹永忠. (2016i). 顯示技術：視覺暫留的應用- 手搖字幕機開發硬體篇. *顯示技術*. Retrieved from http://www.techbang.com/posts/41880

曹永忠, 許智诚, & 蔡英德. (2014a). *Arduino 互动字幕机设计: Using Arduino to Control a Color Led Display with An Android Apps*. 台湾、彰化: 渥瑪數位有限公司.

曹永忠, 許智诚, & 蔡英德. (2014b). *Arduino 光立体魔术方块开发: Using Arduino to Develop a 4* 4 Led Cube based on Persistence of Vision*. 台湾、彰化: 渥瑪數位有限公司.

曹永忠, 許智誠, & 蔡英德. (2014a). *Arduino 互動字幕機設計: The Interaction Design of a Led Display by Arduino Technology* (初版 ed.). 台灣、彰化: 渥瑪數位有限公司.

曹永忠, 許智誠, & 蔡英德. (2014b). *Arduino 手搖字幕機開發:The Development of a Magic-led-display based on Persistence of Vision* (初版 ed.). 台灣、彰化: 渥瑪數位有限公司.

曹永忠, 許智誠, & 蔡英德. (2014c). *Arduino 手搖字幕机开发: Using Arduino to Develop a Led Display of Persistence of Vision*. 台湾、彰化: 渥瑪數位有限公司.

曹永忠, 許智誠, & 蔡英德. (2014d). *Arduino 光立體魔術方塊開發:The*

Development of a 4 * 4 Led Cube based on Persistence of Vision (初版 ed.). 台灣、彰化: 渥瑪數位有限公司.

曹永忠, 許智誠, & 蔡英德. (2014e). *Arduino 旋转字幕机开发: Using Arduino to Develop a Propeller-led-display based on Persistence of Vision*. 台湾、彰化: 渥瑪數位有限公司.

曹永忠, 許智誠, & 蔡英德. (2014f). *Arduino 旋轉字幕機開發: The Development of a Propeller-led-display based on Persistence of Vision*. 台灣、彰化: 渥瑪數位有限公司.

曹永忠, 許智誠, & 蔡英德. (2015a). *86Duino 程式教學(網路通訊篇):86duino Programming (Networking Communication)* (初版 ed.). 台湾、彰化: 渥瑪數位有限公司.

曹永忠, 許智誠, & 蔡英德. (2015b). *86Duino 编程教学(无线通讯篇):86duino Programming (Networking Communication)* (初版 ed.). 台灣、彰化: 渥瑪數位有限公司.

曹永忠, 許智誠, & 蔡英德. (2015c). *Ameba 空气粒子感测装置设计与开发(MQTT 篇):Using Ameba to Develop a PM 2.5 Monitoring Device to MQTT* (初版 ed.). 台湾、彰化: 渥瑪數位有限公司.

曹永忠, 許智誠, & 蔡英德. (2015d). *Ameba 空氣粒子感測裝置設計與開發(MQTT 篇)):Using Ameba to Develop a PM 2.5 Monitoring Device to MQTT* (初版 ed.). 台湾、彰化: 渥瑪數位有限公司.

曹永忠, 許智誠, & 蔡英德. (2015e). *Arduino 实作布手环:Using Arduino to Implementation a Mr. Bu Bracelet* (初版 ed.). 台湾、彰化: 渥瑪數位有限公司.

曹永忠, 許智誠, & 蔡英德. (2015f). *Arduino 程式教學(入門篇):Arduino Programming (Basic Skills & Tricks)* (初版 ed.). 台湾、彰化: 渥玛数位有限公司.

曹永忠, 許智誠, & 蔡英德. (2015g). *Arduino 程式教學(常用模組篇):Arduino Programming (37 Sensor Modules)* (初版 ed.). 台湾、彰化: 渥玛数位有限公司.

曹永忠, 許智誠, & 蔡英德. (2015h). *Arduino 程式教學(無線通訊篇):Arduino Programming (Wireless Communication)* (初版 ed.). 台湾、彰化: 渥瑪數位有限公司.

曹永忠, 許智誠, & 蔡英德. (2015i). *Arduino 编程教学(无线通讯篇):Arduino Programming (Wireless Communication)* (初版 ed.). 台湾、彰化: 渥瑪數位有限公司.

曹永忠, 許智誠, & 蔡英德. (2015j). *Arduino 编程教学(常用模块篇):Arduino Programming (37 Sensor Modules)* (初版 ed.). 台湾、彰化: 渥玛数位有限公司.

曹永忠, 許智誠, & 蔡英德. (2015k). *Arduino 實作布手環:Using Arduino*

to Implementation a Mr. Bu Bracelet (初版 ed.). 台湾、彰化: 渥瑪數位有限公司.

曹永忠, 許智誠, & 蔡英德. (2015l). *Arduino 編程教学(入门篇):Arduino Programming (Basic Skills & Tricks)* (初版 ed.). 台湾、彰化: 渥玛数位有限公司.

曹永忠, 許碩芳, 許智誠, & 蔡英德. (2015a). *Arduino 程式教學(RFID 模組篇):Arduino Programming (RFID Sensors Kit)* (初版 ed.). 台湾、彰化: 渥瑪數位有限公司.

曹永忠, 許碩芳, 許智誠, & 蔡英德. (2015b). *Arduino 編程教学(RFID 模块篇):Arduino Programming (RFID Sensors Kit)* (初版 ed.). 台湾、彰化: 渥瑪數位有限公司.

趙英傑. (2013). *超圖解 Arduino 互動設計入門*. 台灣: 旗標.

趙英傑. (2014). *超圖解 Arduino 互動設計入門(第二版)*. 台灣: 旗標.

Arduino 程式教學 (顯示模組篇)
Arduino Programming (Display Modules)

作　　者：曹永忠、許智誠、蔡英德

發 行 人：黃振庭

出 版 者：崧燁文化事業有限公司

發 行 者：崧燁文化事業有限公司

E-mail：sonbookservice@gmail.com

粉 絲 頁：https://www.facebook.com/
　　　　　sonbookss/

網　　址：https://sonbook.net/

地　　址：台北市中正區重慶南路一段六十一號八
　　　　　樓 815 室

Rm. 815, 8F., No.61, Sec. 1, Chongqing S. Rd.,
Zhongzheng Dist., Taipei City 100, Taiwan

電　　話：(02) 2370-3310

傳　　真：(02) 2388-1990

印　　刷：京峯彩色印刷有限公司 (京峰數位)

律師顧問：廣華律師事務所 張珮琦律師

國家圖書館出版品預行編目資料

Arduino 程式教學 . 顯示模組篇 =
Arduino programming(display
modules) / 曹永忠 , 許智誠 , 蔡英
德著 . -- 第一版 . -- 臺北市 : 崧燁
文化事業有限公司 , 2022.03
　　面；　公分
POD 版
ISBN 978-626-332-080-2(平裝)
1.CST: 微電腦 2.CST: 電腦程式語
言
471.516 111001398

官網

臉書

定　　價：620 元

發行日期：2022 年 03 月第一版

◎本書以 POD 印製